Happy Birthday, Viva

Love

Stacey. 1994

SAVING GRACES

BOOKS BY ROGER B. SWAIN

Earthly Pleasures
Tales from a Biologist's Garden

Field Days
Journal of an Itinerant Biologist

The Practical Gardener
A Guide to Breaking New Ground

Saving Graces
Sojourns of a Backyard Biologist

Saving Graces

SOJOURNS OF A
BACKYARD
BIOLOGIST

ROGER B. SWAIN

Illustrated by Abigail Rorer

Little, Brown and Company
Boston Toronto London

FIRST EDITION

The essays in this book previously appeared in the following publications: "Dime-Store Turtles," *National Public Radio Broadcast;* "Full Pockets" and "The Onion Braider," *New York Times Magazine;* "Blue Birdfood," *Living Bird Quarterly;* "Hand to Tooth Work" and "Dead Stock," *Harrowsmith.* All the remaining essays appeared in the *Audubon Calendar,* except "The Boathouse," which is published here for the first time.

LIBRARY OF CONGRESS CATALOGING-IN-PUBLICATION DATA

Swain, Roger B.
Saving graces: sojourns of a backyard biologist / by Roger B. Swain.—1st ed.
p. cm.
1. Natural history. I. Title.
QH81.S883 1991
508—dc20 91-7767
ISBN 0-316-82471-2 CIP

10 9 8 7 6 5 4 3 2 1

Designed by Barbara Werden

BP

Published simultaneously in Canada by Little, Brown & Company (Canada) Limited

PRINTED IN THE UNITED STATES OF AMERICA

FOR MARGUERITE

whose sterling ingenuity is exceeded
only by her generosity

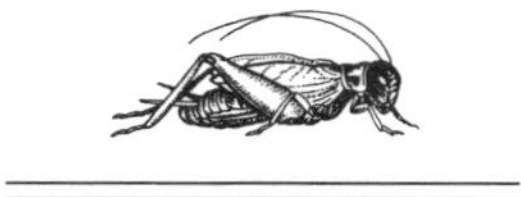

CONTENTS

CONTENTS

"The more the wonders of the visible world
become inaccessible, the more intensely
do its curiosities affect us."

COLETTE

SAVING GRACES

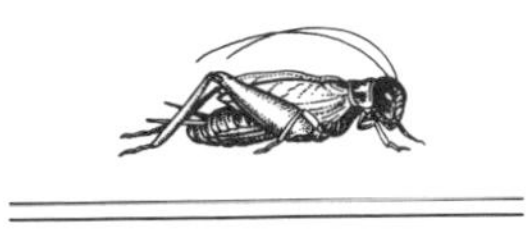

HORSE CHESTNUTS

In my suit-jacket pocket are a couple of horse chestnuts, picked up months ago on a back street. The asphalt was littered with spiny husks and the remains of nuts that cars had run over, but in the gutter I found two burrs that were still intact, inside each a pearl of oiled and polished mahogany.

Horse chestnuts have always been worth having. As children we filled empty shoeboxes with them, or our top bureau drawers. What prompted the annual hoarding was not their taste. Some of us tried repeatedly to eat the hard white meat and were beaten back by bitterness every time. Nor were horse chestnuts very useful. Drilling holes in them and tying them on opposite ends of pieces of string turned them into bolas, useful mostly for decorating telephone wires. Finally, while horse chestnuts never actually rotted,

they quickly lost their initial luster, shrinking and wrinkling until they resembled the dried litchis that Chinese restaurants served for dessert. None of this, however, made horse chestnuts any less desirable, for none of us thought to question their worth. Every September I scaled the horse chestnut in the church parking lot, raised up its rough trunk by faith, pure and simple.

I have outlived both the church and the tree—one burned, the other sawed down—but I still collect horse chestnuts. My pulse no longer quickens at the prospect of gathering a whole half bushel of them, and my top bureau drawer holds nothing but under wear and socks. Yet each fall, whenever I find the curled, yellow, palmate leaves and the spiny husks lying on a sidewalk, I stop to find a couple of unblemished nuts, round ones perhaps, or a pair flattened on one side from having shared a burr with a sibling. Pocketed, they travel with me.

Thirty thousand feet in the air, in a fragile cocoon of aluminum and plastic, a briefcase full of papers wedged in under my feet, it is easy to think globally. The headlines of today's paper folded in the seat pocket in front of me talk of war, acid rain, and impending energy shortages. Down below are chemical dumps leaking, tropical forests burning, whales dying, humans being born. All around me sit experts, their attaché cases filled with reports, charts, bar graphs, and color transparencies.

But in my pocket are two horse chestnuts. I touch them and I am under the back porch, feeding ant lions in their conical pits. One touch and I have a raft on wheels and a matched set of kitchen strainers to seine for water striders and tadpoles. One touch and it is summer vacation and I am going to grow up to be a taxidermist. In whatever stratosphere of world issues I find myself, the horse chestnuts bring me back to earth.

Horse chestnuts will not work for everyone. But other grown-ups, I notice, have their equivalents: the tail feathers of a red-shouldered hawk, a glass bottle filled with beach sand, a lump of copper ore that doubles as a paperweight. Each reminds someone of a place and time when, whether they knew it or not, they had both feet firmly on the ground, a reminder that is the most subtle, and yet the strongest, form of encouragement. Horse chestnuts are my talisman. In the rarefied atmosphere of world responsibility, I find that they work a simple magic, reminding me what it is exactly that I have grown up to care for.

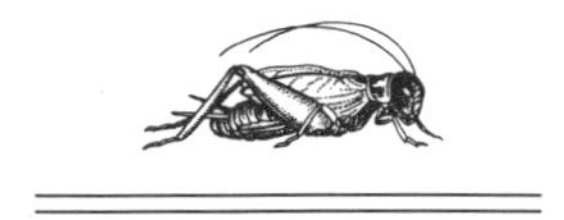

HERMIT CRABS

A low tide at noon has calmed the bay, where sandpipers rest on a spur of damp beach. A herring gull pecking at something backs off reluctantly at my approach. Bucketless, I am collecting shells in the crown of my hat, discarding earlier specimens when I find something better. A fawn-skinned surf clam, a quahog with its spot of wampum purple, a razor clam, a slipper shell, a bay scallop, and a handful of translucent yellow-and-orange jingle shells fill my cap. I have also saved a black and leathery egg case of a skate, a necklace of whelk eggs, and a lump of sand-scoured green glass.

In the hot sun, it feels good to walk barefoot through the tide pools on the flats, treading the small sand ridges sculpted by the waves. What I can't see beneath the water's glare, I sense with my toes, here the slickness of brown kelp, the mush of sea lettuce,

there the hardness of shell. Stooping, I retrieve a baby-faced moon shell that feels too heavy to be empty. As I turn it over I expect an operculum sealing the entrance and find instead a cluster of claws, the massed legs of a hermit crab.

Hand-me-down housing is a way of life for these crustaceans. Unlike their armored lobster and shrimp kin, hermit crabs have soft, vulnerable abdomens, designed to fit neatly into the right-handed spirals of empty mollusk shells. I could not easily dislodge the one I hold now; its hindmost appendages are tightly gripping the moon shell's inner walls. But in a few moments the forelegs stir, the stalked eyes appear, and the hermit crab scrabbles against my hand.

Fish can swallow hermit crabs, house and all. Those hermit crabs occupying smaller periwinkle shells frequently sport a pinkish crust, a hydrozoan called snail fur. The hydrozoan rides piggyback to new waters, and its stinging cells may save them both from attack. Far more basic to the hermit crab's survival, however, is simply finding a shell that fits.

Judging from the modest size of this particular moon shell, the hermit crab inside is a year or two old and still growing. To do so it will have to find additional shells, each of which must be tight enough to hold on to, and roomy enough for the animal to put on weight. Large, intact, empty gastropod shells are few and far between. Each prospective house this hermit crab encounters it will examine closely, measuring the shell with its claws, and then try on for fit in a lightning fast shift of quarters. Should the new shell prove superior, the one it leaves behind will soon be occupied by a slightly smaller hermit crab, whose now empty shell will be taken by a smaller hermit crab still, and so on down the line.

This shell shuffling of hermit crabs is an example of a vacancy chain, a term originally coined to describe our own behavior, as in the allocation of parishes among Episcopal priests. Now, at a time when so many humans are homeless, comparisons between our lot and that of hermit crabs seem especially apt.

Releasing the hermit crab and watching it scurry off across the sandy bottom toward its future, I would not wish on us the equivalent of its endless relocations. But we could do with a much better domestic fit than whole families living in one room or individuals sleeping on the street. Shelter for all is within our grasp, for we have the tools and skills to build and repair and redesign. We are shell collectors, but not hermits. To share our roof with others is the gift of a permanent home.

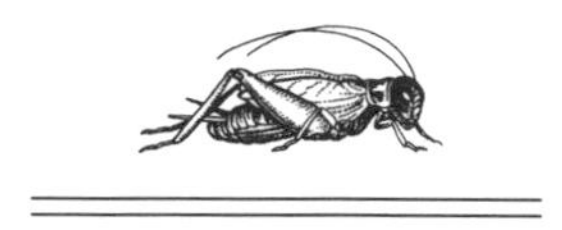

WASPS AND WOLVES

Gray wolves once howled on Atlantic beaches. Colonists, huddled in huts on the edge of the New World, listened to the chorus of wind, wave, and wolf and were afraid. On November 9, 1630, scarcely a decade after the *Mayflower* dropped anchor, the governors of Massachusetts passed the first wolf bounty law in America. Up and down the coast, wherever people saw the huge footprints in the sand, they remembered stories of a diabolical beast, red-eyed and slavering, and they put a price on its head. With muskets and wolfhounds, with spring-loaded tallow balls and deep pits, the shore was ridded of wolves.

When the settlers moved inland, they carried their fears with them. Every wolf was an unbearable threat to livestock, family, and future. Wagons rolled west, and the killing intensified. A wave of strychnine washed across the Great Plains in the 1880s, erasing not only the footprints of wolves but those of every animal attracted to the poisoned meat. Later, government hunters replaced bounty hunters, and sportsmen took to the sky, shooting whole packs at once from low-flying aircraft. But the savage slaughter could not continue. There weren't enough wolves.

To hear wild wolves now, one must go to northernmost Minnesota or Michigan's Upper Peninsula, to Glacier National Park in Montana or Isle Royale in Lake Superior. Farther north, in Canada and Alaska, there are more wolves. Farther north, there are fewer people.

Back where it all began, in the heavily populated towns of the East Coast, wolves have been relegated to television specials and wildlife stamps. On a sunny summer afternoon more attention is devoted to hamburgers cooking on a backyard grill, the potato salad, the cold beer. No one worries about predators. But no sooner are the paper plates filled and served than yellow jackets appear, half a dozen or more flying huntresses. Some land and begin carving chunks from hamburgers, others disappear into opened cans of beer and soda. You can't shoo a yellow jacket away, but you can provoke her into stinging. She is armed with a reusable stinger longer and more painful than a honeybee's, which she will use if she finds herself pinned under a hand or accidentally washed into someone's mouth.

Receiving one sting is painful enough; more, multiply so. People who are allergic to the venom must be especially cautious, and everyone has heard stories of innocent people being fatally stung, terrible tales told and retold. Small wonder that people are quick to try to kill all the yellow jackets that appear, bashing them one at a time on the picnic table with more force than necessary, lest they sting. This seldom clears the air, though, for the dead are quickly replaced by yellow jackets from a nest somewhere that may contain as many as two thousand. The nest is well hidden, being underground. The best way to find it is by accident, by stepping on the entrance or driving a power mower over it. Then a swarm of defenders attacks, giving the victim additional incentive, when the swellings have subsided, to return after dark and pour poison down the entrance hole.

Yellow jacket nests are hard to find, but the nests of other common wasps are conspicuous. Bald-faced hornets build a nest much like a yellow jacket's, but instead of hiding it in an excavated chamber underground, they suspend it from a tree branch or the eave of a house, a great gray-paper globe. Eaves are also a favorite nesting site for Polistes wasps, builders of the single combs that hang horizontally without any protective covering. Regardless of what the nest looks like, its occupants sting, and for many people that is reason enough to burn or spray or knock it down. The fewer wasps around the place, the less likely anyone is to get hurt.

That, of course, is what we used to say about wolves. Most of us don't say it anymore, for our feelings about wolves have changed. Biologists and others who have spent years with the animals tell us that we could learn much about ourselves by studying wolf society. They tell us that we had no reason to be so afraid, that wolves weren't what we imagined they were. Many of us agree, and feel ashamed of the killing we once condoned.

Of course, it is easier to say good things about wolves now that no one we know is losing sheep. Put another way, wildflowers are more beautiful if they aren't weeds in your own field. Farmers who tried to grow crops in land thick with orange hawkweed and Queen Anne's lace called them devil's paintbrush and devil's plague. Dandelion seed heads are fragile beauties—in photographs. But even if wolves proved to be less noble at close hand, it would be worth it to see one now and then: a wild one, not one in some zoo. Yet who can imagine a wolf sitting at the edge of the tennis court, trotting across the parking lot, stopping to drink from the wading pool?

There will never be wolves on Main Street. But there are wasps, social hunters on a different scale. If we can forget the few times we were stung, ignore the fearful warnings of friends, we can watch wasps catching flies and small caterpillars to feed their young. We can watch as they scrape up wood fibers into pulpy balls to carry back and add to the nest. In the fall, when all the sterile workers and the males are dead, and when next year's queens have gone off to hibernate under bark and loose shingles, we can cut the big bald-faced hornets' nest out of the lilac. Slicing open its many-layered paper envelope, we will find level upon level of comb, intricate architecture built without blueprints or a foreman.

Wolves howl in the boreal forests, but few of us will ever hear them. Wasps, on the other hand, still come to every picnic. Make room for them. We shouldn't have to enjoy wilderness at a distance.

WILD EYES

When I was young, someone discovered a mastodon tusk less than a quarter mile from my house. It lay in waist-deep water on the bottom of a shallow pond where we went to catch sunfish and turtles. The six-and-a-half-foot-long tusk, with its brown and corroded exterior, looked more like a section of rotting log than a fossil tooth. Only when it was sliced open by the museum that put it on display could you see the concentric rings of ivory in its interior. For a while, the discoverer was a local celebrity. But he wasn't all that special, just luckier. If another one of us had snagged our fishing line on the tusk and waded into the green, murky water to rescue our hook, then we would have had our picture in the paper instead.

I went through childhood fully expecting to discover a mastodon tusk myself. I knew they existed, and the only obstacle I could see was recognizing the fossil when I found it. The news that someone had come upon a 42,000-year-old specimen amidst beer bottles and sunken shopping carts was further proof that such discoveries were to be expected in daily life. Ancient teeth, of course, weren't the only things I expected to run across. I was also anticipating dinosaur footprints, trilobites, and Indian arrowheads. I hung around construction crews, watching what their backhoes dug up. I inspected every rock's dimples for signs of prehistoric life. Walking to the library or back from the barbershop, I pocketed every stone that looked remotely like a pre-Columbian tool.

While I was at it, I checked the base of all the big pine trees in the neighborhood for owl pellets, the disgorged balls of hair and bone that could be pulled apart to see what the bird had eaten. I saved any feather that might have come from a hawk. I painted the trunk of the sugar maple by the sidewalk in front of our house with a mixture of stale beer, brown sugar, and rotten strawberries, in hope of attracting green-winged luna moths. And when I visited the park up the street, I invariably checked the little cave there beneath a rock ledge, in case a porcupine or fox had taken up residence.

I never found any porcupines or foxes, and the feathers mostly came from pigeons. The pines yielded

pop bottles, not owl pellets, though each of these provided a two-cent consolation. And if prehistoric man ever used any of the stones I brought home, he threw them at something. In retrospect, this scarcity of wild things, dead or alive, was not surprising. I grew up on the outskirts of a big city, where the lots were small, the streets paved, and every house stocked with an assortment of children, at least some of whom were looking for the same things as I. A skull on the sidewalk of our town could be expected to remain there about as long as a five-dollar bill.

What surprises me now, looking back, is how tolerant parents were of our wild-eyed expectations. They never pointed out how useless it was to look for fossils in granite. Nor did they reveal the true scarcity of owls and Indian artifacts. Instead they gave us full license to search. They equipped us with field guides and containers, and housed our motley collections of might-be and could-have-beens.

Now and then we did come up with something, a praying mantis, a garter snake, a baby raccoon. I can see now that these successes pleased our parents as much as they pleased us. Not just because we had distinguished ourselves as naturalists, but because we had beaten the odds. Our parents took vicarious pleasure in our finds, remembering, I think, what it felt like to be in our sneakers.

All of us are born wild-eyed, but our outlook changes with time. First you stop seeing camels in the clouds, then you outgrow the fear that there is a snake in your bed, then you learn that there aren't really alligators in the sewers. Plans to become a forest ranger disappear along with any chance of seeing an ivory-billed woodpecker. The expansive optimism of childhood becomes as limited as a two-week summer vacation.

It never disappears completely, of course. Childhood leaves its mark on every adult. I still look at a pond and assume that there are big fish to be caught there. I look at tracks in the snow and think they must have been made by something a lot bigger than a dog. Now and then I am vindicated. Wildlife shows up unannounced inside civilization's borders—a moose is spotted in someone's backyard, a bear appears on the median strip of a turnpike, a whale blunders into a congested harbor. Every one of these occurrences recalls me from my wise cynicism about the declining state of wilderness. I stop and remind myself how I used to look at the world. I reach for what has become my favorite motto: "You are only young once, but you are never too old to be immature."

I was tempted to tell my children that they didn't have a chance of catching a mouse in their bedroom with the lights on. But I showed them how to set the box trap anyway. Ten minutes later they were back with a mouse that had come out of the woodwork, obligingly walked to the center of the room and into their trap. Their mouse, our mouse, greater than any mastodon tusk. Together we carefully carried it down the snowy street, and let it go safely in someone's warm garage.

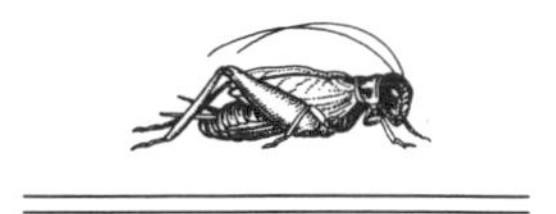

BRIGHT STARS, DARK NIGHTS

Astronomers were the first to sound the alarm. Observatories in university towns, blinded by night light, were relegated to teaching and not to research. Bigger, stronger telescopes were built on desert mountaintops in a climate of clearer, darker skies. But cities continued their relentless advance, and now these lenses, too, are threatened by the approaching nocturnal glow.

The stargazers will not surrender without a fight. In some quarters they have moved to stop development near observatories; in others, the campaign is simply to require reflectors over streetlights and to change the bulbs from high-pressure sodium or mercury vapor to low-pressure sodium with its narrower spectrum of illumination. But there is general agreement that the future for optical telescopes is in space, where the sky is always three times darker than any place on Earth.

For most of us the nights are not what they used to be. Sunset near human habitation is followed not by the Stygian darkness of antiquity but by an artificial dusk that is turned off only by dawn. No one disputes that this constant illumination allows us to lengthen our work and play, or that it cuts down on accidents and crime. But in our efforts to banish shadows we have created bright deserts. What bleaker instance than an empty, night-lit parking lot?

Our own species shies from the dark, but legions of others—from fireflies to whippoorwills—depend on it. Plants and animals rely on changes in day length to determine when to bloom, lay eggs, molt, or migrate. Even our own bodies have internal clocks sensitive to light and dark. Before we sacrifice more kilowatts toward universal illumination, let us remember that we have a dark-adapted heritage, too.

Outdoors on a summer evening in rural New Hampshire I watch the bats come out from their roosts under the eaves as dusk falls—fast, black butterflies. As the sky darkens my eyes compensate, my vision improving until it is some ten thousand times more sensitive. Even with no moon I can make out the ash trees, the stone wall, the cats hunting on the lawn, though the picture has shifted from color to old-fashioned black and white. The rasping of crickets surges and ebbs, punctuated by the deeper croak-

ing of frogs. Far off in the valley a truck rumbles past. I smell the world cooling off, like a pie newly taken from the oven.

Overhead, fair weather and darkness have produced a sky more scintillating than any fireworks display. Splashed in a swath from one horizon to the other is the glistening mass of our galaxy. It is a sight that always stuns city-dwellers who, growing up in luminous smudge, know the Milky Way only by its name. No one can look on these billions and billions of stars and not be moved. The sheer scale of the galaxy forces the observer, like Ralph Waldo Emerson, to feel the stars "pouring satire upon the pompous business of the day."

Humility is not the sole lesson, only the first. The heavens hold many keys. Careful observation can unlock the secrets of the seasons, the rhythm of the tides, the timing of eclipses—all the inventions of astronomy. Most of us look up at all those stars and are overwhelmed, but in every civilization there have been stargazers who have independently discovered celestial order in the universe. The best and the brightest have never been afraid of the dark.

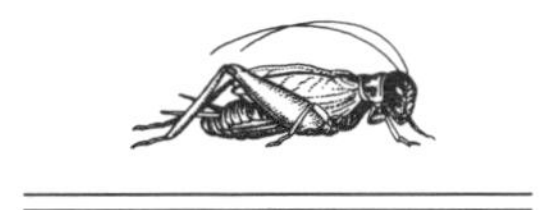

TO SMELL

It was a game we played as teenagers on August nights—seeing who could bicycle home in the dark with no flashlight. The bravest even pedaled off into the shadows with their hands up in the air. At midnight, on the long narrow country road—half tar, half dirt—it was not automobiles or walkers that we risked running into but the unforgiving trunks of paper birches and fir balsams. By far the most common misfortune, however, was simply veering off the pavement and sinking into the soft earth of the ditch. But what was risky at first became an easy stunt as the weeks rolled by, most of us learning to ride the thin ribbon of warm tar and packed sand to our respective beds.

On clear nights, if you looked straight up you could steer by the stars. It was easier, though, to navigate with your nose. If you breathed deeply, sucking the air in through your nostrils as your knees pumped up and down, you could tell where the trees were by their smell, by the scent of their needles and the pitch oozing out of their trunks and the accompanying earthy scents of mushrooms, lichens, and moss. You could sense the edge of fields, a change to the aroma of crushed hay-scented ferns or grass drying on a freshly mown softball diamond. A thick haze of moist musk marked the low, swampy spots as clearly as the sounds of frogs. And while individual driveways were easy to miss, there was no missing the houses themselves. In this summer community in the northern woods, even in mid-August wood smoke poured from every chimney.

Though our noses kept us on the straight and narrow those summer nights, none of us ever mentioned it at the time. Unless you are Cyrano de Bergerac or Jimmy Durante, one's nose isn't a topic for conversation. So many people are sensitive about this portion of their anatomy that it isn't polite even to mention the subject. But why should we be so sensitive? Surely it isn't trivial variations in shape. No, I suspect our reserve has more to do with the knowledge that our nose is not very sensitive, that compared with other animals' noses it is not really up to snuff. Dogs can track a man across hard ground and distinguish the smell of his footsteps from those of another. Salmon can find the stream in which they hatched by the odor of the water. Male moths can detect a female at a distance of a million times her own body length. In such company it is easy to feel inferior.

But let us not be defensive. Our noses function quite well. An average run-of-the-face nose can recognize a great number of different odors. Some are detectable even in tiny amounts, such as the scent of vanilla or the trace of the ethyl mercaptan in natural gas. The gourmet cook's vaunted sense of taste is really a practiced sense of smell. Anyone who can weigh the merits of two cups of tea or wine or olive oil has trained his nose. For such an individual this sense has not so much been lost as it has been miscataloged.

For the rest of us our sense of smell is too often simply sidetracked. Living in the fast lane means being chronically short of both time and breath. When you have to gulp your air, there is no time to savor it, no time to let it linger in the nose. You might as well try to listen to music wearing earmuffs or look at art from behind dark glasses. We live in a world whose odors are as rich and varied as its sights and sounds, but those who rush through it end up smelling nothing beyond their own sweat. To slow down is to discover that, in more ways than one, you smell better.

Stand still a moment and hold your nose up. You don't need a weathervane to tell which way the wind is blowing. As the wind swings around it brings first the smell of salt marshes, then smokestacks, farmland, woods, then the sea again. There is a soft smell that signals the coming of rain; the clean penetrating odor of ozone tells you where lightning has been.

We all know about the fragrances of flowers, but blossoms are not the only plant parts that smell as sweet. There are the bruised leaves of pennyroyal and pine and eucalyptus; the fruit of fox grapes, wild strawberries, ripe quince; the twigs of spicebush and sassafras; the heartwood of red cedar. At first every odor is a new one; but then patterns begin to emerge as new scents remind us of old ones. No one has ever devised a precise way to define a scent and so the world is filled with observations that something smells like something-or-other else. But the connections are often justified, for the number of aromatic compounds is finite; chewing either a twig of sweet birch or the evergreen leaves of checkerberry yields the same sensation: methyl salicylate, or oil of wintergreen.

Some animal scents are inescapable, like the smell of a skunk or the rank odor of woods through which a herd of peccaries has recently passed. Others take a bit of searching, such as the leathery odor of castoreum left by beavers on the small piles of mud they heap up to mark their territory. If you smell where a honeybee stings you, you will detect banana oil. Turn over enough rocks and you will come across a colony of the tiny yellow ants that smell of citronella.

Whatever the scent, all have one thing in common that distinguishes them from all other sensations: none of them can be recalled at will. Melodies can be stored in the mind and hummed later; scenes can be sketched from memory. The smell of sweet fern or beeswax or wood smoke, on the other hand, can only be enjoyed at the moment.

But just because a scent cannot be recalled does not mean it is forgotten. Smells, in fact, are the most powerful releasers of memory. Sometimes it takes only a single aroma, a slight whiff of something once encountered, to bring back entirely a moment of one's past. The chance to grasp the brass ring of reincarnation is more than enough reason for me to be pleased with my nose. I have only to crush fresh fir-balsam needles and sniff their boreal fragrance and I am a teenager again, laughing in the darkness, overflowing with self-confidence, careening happily toward home.

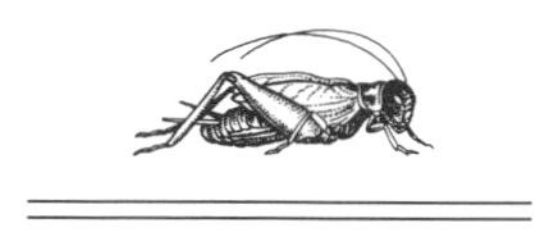

TREE DREAMS

I sleep in the woods. Straight trunks of sugar maple hold me between earth and sky. Stout limbs cradle and protect me. Though I spend many unconscious hours here in the dark, I am never afraid. For when I awake, I awake in bed.

This bed would have pleased Odysseus and Penelope. Though it has no living tree as a corner post, it is as immovable as theirs, so massive are its uprights, so great its headboard. It is a bed of substance, and of years. The doctor who delivered my grandmother slept here, on rope springs and a mattress stuffed with straw.

A grove of sugar maple trees, to a logger, is made up of sawlogs and cordwood just waiting to be felled, bucked to length, and yarded. A sawyer looking over these logs arriving at the mill sees boards and planks and timbers neatly stacked under the bark. Because they look at the trees this way, both are sometimes told that they do not really see the forest, that they should be less quick to see the trees as furniture, flooring, and breadboards.

But who of us is so free of self-interest as to have never looked at a blueberry bush and thought muffins, nor played in a stream and tried to harness it with a small dam? All of us translate the things we see into forms we are familiar with. It is natural, and useful. And what the purists do not realize is that the skill of translating works in reverse.

I, too, can look at a sugar maple tree and see syrup, cords of firewood, and clear lumber. But in a parallel manner I can look at the bed I am sleeping in and put it back into a tree somewhere in the southern Appalachians. By counting the annual rings on the headboard, I know that date. It is two hundred years ago.

I am awakened by a flock of Carolina parakeets that come screaming out of the forest and settle on a nearby tree, carpeting it in orange, yellow, and green. Below me a buffalo cow with calf scratches herself on the tree's bark. Wolves and cougars prowl these woods. Migrating passenger pigeons blot out the sun. John James Audubon is one year old.

If I were a purist, and resolutely kept trees and fur-

niture separate in my mind, I would miss this. Most of my life would be spent enmeshed in civilization. Wilderness would be reserved for special occasions. But by simply recognizing the wild origins of domestic life, I am permitted almost unlimited woodland wanderings, tramps across marshes, views from high bluffs. Almost anything made of wood can be grounds for an excursion. Cedar pencils, wooden clothespins, excelsior. These simple household objects contain volumes of natural history. Provided that they are read in translation.

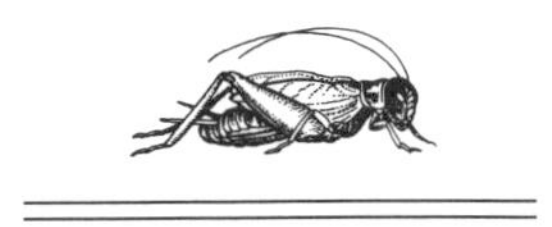

WEATHER

In our judgment of the weather, we have become exceptionally narrow-minded. The forecaster on the radio this morning issued a "winter storm warning," and if the predicted flakes materialize, a couple of inches of accumulation will create a state of "snow emergency." Last summer this same forecaster persisted in describing possible showers, in the middle of a drought, as a "threat of rain." The only precipitation that is greeted with any enthusiasm these days is snow on Christmas Eve. Otherwise, "good" weather means sunny, mild, and dry.

If the issue were to appear on a ballot, the majority of the nation's populace would, I suspect, vote to severely restrict rain, mist, fog, hail, sleet, snow, and ice. High winds would certainly be outlawed. I can imagine minimum temperatures being set state by state, figures roughly equal to the highway speed limits in effect before the price of oil went up. And local ordinances allowing the presence of a few clouds of the fair-weather sort, and now and then a gentle breeze or heavy dew.

At present, however, the weather remains outside our regulatory discretion. When the skies cloud over and the mercury falls, we can either move indoors or move away. Each year more people join the geese and head south, some for the winter, some for good. The ones who stay behind will spend the winter lamenting the cost of home heating. By spring we will be suffering to varying extents from cabin fever, the result of being cooped up too long.

Outside the kitchen window, chickadees are pulling sunflower seeds from the feeder. Part of the entertainment for us is the marvel of such small bundles of feathers surviving the weather. But is the weather all that bad? I think we have set our thermostats too high. In Tierra del Fuego, midsummer temperatures average 45 degrees by day, and yet Charles Darwin noted there in his journal of December 25, 1832, that he had seen a woman nursing a newborn baby, both mother and child apparently oblivious to the sleet that fell and thawed on their naked bodies. A month later he wrote that while he and others of his party huddled around a fire for warmth, undressed Fuegians positioned at some distance were "streaming with perspiration at undergoing such a roasting."

This tolerance of cold comes from long acclimation. It is not something we can acquire overnight. But our intolerance is also the fruit of years of conditioning. Each of us has been taught to believe that we could never survive such exposure, that wet feet and cold hands are sure to bring on illness—a "cold in the nose" or a "cold in the head." As children we knew better. We spent all day out sledding, or ran about in the rain, ignoring our mother's warnings that we would catch double pneumonia. And we survived.

Some weather is undeniably bad for you. We do well to avoid hurricanes, tornadoes, and electrical storms. But we should not balk at rain and snow. If the day turns out to be cold or wet we don't have to cancel our plans to spend it outdoors. With an extra wool sweater or two, and some boot grease in winter, a broad-brimmed hat and a slicker in summer, even those most accustomed to the shelter of climate-controlled buildings can return to their youth and safely venture outside.

Raindrops falling on the open water of a lily pond, salamanders crossing a path on a rainy night. The cold squeak of snow underfoot, sea smoke in the bay on a frigid morning. Once in a great while, an entire grove of spruce transformed by ice into a giant crystal chandelier. These are the transitory beauties of bad weather, the bounty of those who are broad-minded.

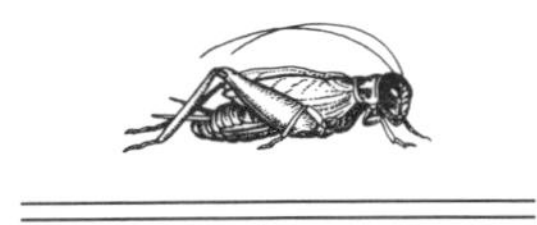

DIME-STORE TURTLES

Of all the things you can't buy anymore, I miss turtles the most. Pet turtles were once as essential to summer as Popsicles and nearly as common. Every five-and-ten-cent store sold them. I remember the ones at the back of the F. W. Woolworth store—back past the hair curlers, the wrapping paper, the clothespins and cosmetics. They were baby turtles, dark green silver dollars, each with a red spot on the side of its head.

I don't remember how much these red-eared pond sliders cost. Enough that you couldn't have one every time you asked, but cheap enough, too, that everybody could take one home sooner or later. You didn't buy just the turtle, of course. You also bought a clear plastic bowl that had an island in the center topped by a two-piece palm tree—brown trunk and green plastic fronds. Someday, these turtle bowls are going to

be collectors' items, and people are going to have to guess what they were once used for, because since 1975 the interstate shipment and sale of baby turtles has been illegal.

One might assume that the legislation was intended to protect the turtles. As pets, most of these baby turtles were short-lived. Had they been long-lived, the market would soon have been saturated. There were rumors about someone's keeping one of these turtles for years and years, but most of us were resigned to having our turtles expire after a few weeks or months. It was as natural as finding goldfish floating belly-up in the bowl.

But it wasn't the well being of the turtles that the law was intended to protect. It was the well-being of their owners. In 1962, baby turtles were discovered to be carriers of salmonella, the bacterium that can cause virulent bouts of diarrhea in humans. In fact, turtles act as a sort of biological sponge for salmonella, soaking it up from the environment and remaining contaminated indefinitely. Children who put their pet turtles in their mouths while pretending to bite off the turtles' heads got salmonella. But you didn't have to mouth your turtle directly to become infected; often all you had to do was change the water in your turtle's bowl. These were not isolated instances. An estimated 280,000 cases a year, or 14 percent of all human salmonella cases in the United States, were attributed to the keeping of pet turtles. Follow-up studies have shown a significant reduction in childhood illness as a result of banning the sale of turtles less than four inches long, the size most often kept as pets.

But what was a step forward in public health has been a step backward in education. How can children who have never handled a turtle appreciate the fable of the tortoise and the hare? Or such expressions as "turtleneck" or "to turn turtle?" More critical, will they grow up to understand the need for wetland conservation without having had the chance to develop a fondness for turtles, one of the more attractive denizens of swamps? In our eagerness to be protected, once again we seem to have cut ourselves off from nature. This rational xenophobia will in the end, I fear, be the greater evil. This isn't a call for a return to the bad old days when we embraced turtles indiscriminately. Just a reminder that we might be better off if, now and then, we stuck our own necks out.

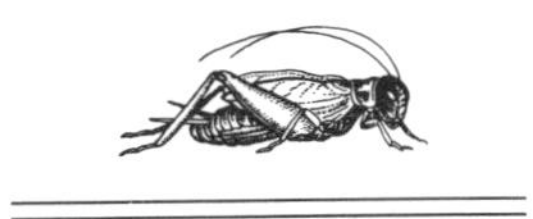

HIGH POINTS

A thin mantle of gritty soil partially covered the rough, gray granite ledges of the low hill and provided roothold for sweet fern, goldenrod, and New England asters. Patches of thin-stemmed grass and blueberry bushes spread across the open summit. To the east, looking out over the tops of the scrub oaks and chokecherries below, you could see the city sprawled in the sunlight and, beyond, the bright water of the harbor and the hazy silhouettes of islands in the distance. A hundred feet overhead a red-tailed hawk regularly circled, its eyes sharp for snakes.

The blueberries, growing in a dense copse underfoot, were the low-bush kind, with the succulent, light-blue fruit. In season whole families and their dogs hiked up on Saturday afternoons, bringing baskets and buckets. Children who were too young to travel under their own power rode on their fathers' backs. Deposited in the middle of a particularly good patch while others picked, many of these youngest got their first taste of the great outdoors. And in the process the hilltop was as indelibly stamped on their psyches as the purple stains on their clothing.

The hill still exists, but its presence is better in memory. Many years ago someone claimed the ground by building a house on the very top. Others followed immediately. The shortage of water and surfeit of rock that had kept away permanent residents for centuries were in the end conquered by iron pipe and dynamite. Surveyors sliced the land into lots, bulldozers made short work of the boulders and the blueberries. A maze of dead-end streets was laid out and lined with contemporary homes, each with its lawn on imported topsoil, each with a slick black drive and multi-car garage. The hawk is gone now, and the people who once picked blueberries under his supervision are none of them young.

"For Sale" signs appear up there from time to time. The view is still a big selling point for the homes, and prospective buyers gaze out over the housetops below, pointing out the sea to one another. But the hill is only a fraction of what it once was and the view has also depreciated.

By now most of us are accustomed to witnessing

local hills engulfed by the human tide. Subdivisions seem to settle on summits like seagulls on pilings. We might grieve more over the loss of our childhood haunts but for the discovery of more distant peaks. By the time the hills of first memory are built over we have moved on to higher, wilder ranges. There we can keep close company with clouds or walk in the preserves of ancient gods or watch as climbers with a sure grip on rock and fear mount the most forbidding pinnacles of all.

Why should we be drawn to heights? Our reasons are elusive and varied. Sir Edmund Hillary was being flippant when he replied to a reporter's question about why he climbed Mount Everest with "Because it is there." If any single cause unites us it is the sanctuary that high points offer, an escape from the leaking gutter, the opinion poll, the leveraged buyout. From the top the world falls away.

This same serenity extends even to those who contemplate mountains from below. "Mountains are the beginning and the end of all natural scenery," wrote John Ruskin, England's first professor of art. They catch the last rays of sunlight, the first snows of winter. Even from a distance they are a tranquil sight, soothing and calming and relaxing us all.

The Japanese, connoisseurs of landscapes, have learned to see snow-capped mountains covered with mist and dotted with small caves in single stones called *suiseki*. Someday we too may have to find solace in such tight quarters. But for now lower powers of magnification suffice. Anyone, young or old, can turn a small hill into a large mountain. All that is required is an empty summit and imagination. With no sure measure of scale beyond variable rocks and trees, distances and heights can be enlarged at will. Ledges become escarpments; crevices, canyons.

What is lost when a ski lift, a microwave tower, or a house is built on a hilltop is this freedom of perspective. With a yardstick to measure things by we are forced to accept the dimensions of reality. Buildings, however utilitarian, don't simply occupy wild ground; they fetter its spirit.

The more concentrated our population, the more precious the remaining unoccupied and unharnessed heights. As a guide to their care we might do well to borrow from technical rock climbers, men and women who climb near-vertical rock faces with the assistance of ropes. Years ago these ropes were clipped to pitons, metal ringbolts hammered into cracks in the rock. In many instances the pitons were left in the cliff, where they served not only as a constant reminder that the face had already been climbed but as a route map as well.

Today climbers practice what they call "clean

climbing," a technique pioneered by the British some twenty-five years ago on some of the more overused mountains in Wales. Instead of being attached with pitons, ropes are threaded through loops or slings that hang from small chunks of metal. These chunks are wedged into crevices where they will bear the climber's weight, but can be removed easily when the last climber on the rope passes by. The result is a rock face as clean as the glaciers left it.

Let us try this ourselves, not just on distant peaks or in national parks, but on those small hills near home, the ones that we grew up on, the ones that we will die looking out upon. We have become masters of topography. We can rearrange the landscape to suit our fancy, and we can build wherever we please. But having learned to move mountains, we should not forget that mountains still have the power to move us.

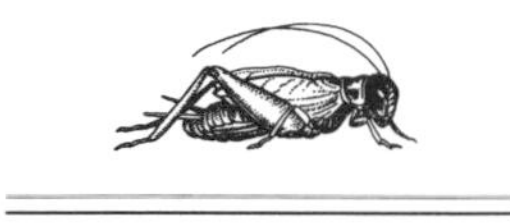

UNTIMELY ESSENCE

Orange-jacketed hunters patrol the surrounding hills. A cold autumn wind, pushing clouds out of the north, sweeps the last leaves from the branches and sinks them against the beaver dam. "No deer," I sing out, my breath a frosty plume. On around the rim of the pond I crunch, pushing through the thickets of bare stems, the curled brown ferns, to the inlet where the brook flows down over rocks and there is always open water.

Snow is in the forecast, the promise of an early winter. And yet a little way upstream, beyond the sharpened stumps of quaking aspen, the witch hazel is in bloom. What benighted madness this is, to flower now. Wrapped in yellow threads, the twigs of the small understory tree fork and arch across the brook. Close up, the lightly scented flowers are grouped in threes, each with four ribbonlike petals scarcely an inch in length that twist and spiral together. The golden, scallop-edged leaves are nearly gone, but there can be no mistaken identity. Of all the trees in eastern North America, witch hazel is last to bloom.

"Snapping-alder" some call this tree. The seeds from last year mature with the new blossoms, and as the fat capsules ripen, their lining contracts until the pair of smooth, black, shiny seeds within pop out like watermelon seeds shot from between a finger and thumb. People who have brought witch hazel branches indoors say the seeds are launched with an audible report and a range of thirty feet.

Dowsers once cut forked branches of witch hazel to search for hidden water and before that American Indians used leaf tea to treat colds, rubbed extracts of the astringent bark on bruises and sore muscles, and added twigs to their sweat baths. When I was growing up, every barber shop had bottles of witch hazel water, the aromatic essential oil distilled from witch hazel brush and mixed with a little alcohol. The barber rubbed it on the back of your neck after a haircut to soothe the chafed skin.

The real essence of the witch hazel, however, is its timing. The unfurling of the flowers this late in the year is no casual appearance, done for the enjoyment of winter walkers. Rather, the witch hazel is out to attract the attention of an equally untimely group of insects. Not butterflies, or bees, or even flies, but a group of winter moths. Some fifty species of owlet moths, in the family Noctuidae, are abroad at night in the dead of winter in northern hardwood forests. Their caterpillars feed on tree buds during early spring, then are quiescent all summer. The adults emerge late in the fall, live out the winter, and die when they have laid their eggs. Among the advantages to this off-season existence is sanctuary from night-hunting birds and bats. Of course, the cold temperatures and food shortages that have cleared the air of these predators are the death of most winged insects as well.

These dull-colored moths—black, brown, gray, or cream—wait out the cold, insulated by thick pile coats and by the layer of fallen leaves under which they hide. But on those sporadic winter nights when the temperature has risen to near freezing, they begin to shiver, vibrating their wings until they have raised their body temperature to the 85 degrees necessary for takeoff. Generally, the moths seek out the sap oozing from injured trunks and branches, but on late-autumn nights they gather at witch hazel, drawn to the nectar on its golden boughs.

Night watches are not for me, not now with family waiting. It is sufficient to have witnessed once more such exquisite eccentricity. The witch hazel and its winter moths belong with fish that fly and birds that don't, with tumbleweeds, water striders, and upside-down sloths. Yet again Nature has set a thing upon its ears in an evolutionary sleight of hand. I applaud without needing to see the moths themselves. It is enough to know how the magic is done, to see where they have fed. In the slate light of late afternoon, I cross the stony brook one last time. Like a deer hunter content to have seen footprints.

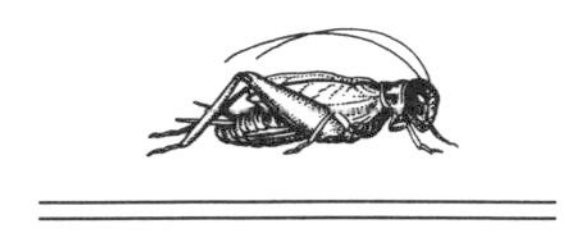

SLOW BOATS

The South American basilisk runs on water. This tropical lizard escapes when disturbed by jumping into forest streams or rivers with a loud splash and racing across the surface on its hind legs. Speed and a fringe of scales on the sides of its toes keep it from sinking.

A powerboat breaking the Saturday-morning calm brings the basilisk to mind. Trailing twin plumes of white froth and two-cycle exhaust, the owner escapes the jungle corporate. A drop in pitch marks the skimming craft's passage. Seconds later it banks around the oxbow and disappears downriver. Only the boat's wake lingers, lifting the lily pads, slapping the banks.

This canoe, by contrast, is going nowhere. Its bow nuzzles a stand of cattails. Off its stern, a bright blue damselfly perches atop a red-and-white bobber moored to a fishing line tied to the end of a cane pole.

Lunch, worms, and binoculars are piled amidships. Within arm's reach, white pond lily blooms perfume the air, attracting honeybees that rummage for pollen in the thickets of their golden stamens. A dozen yards off, three Canada geese glide by, watching us. Under the hull are sunfish, nesting in dinner-plate-size depressions on the sandy bottom.

"Believe me, my young friend, there is *nothing*—absolutely nothing—half so much worth doing as simply messing about in boats," said Rat to Mole in *The Wind in the Willows*.

We are all drawn to the banks of rivers, to the shores of ponds and lakes. But lacking gills, fins, webbed feet, or a sculling tail, we humans are singularly ill-equipped to be aquatic. We roll our pants, we dip our fingers, we cast a line. Sometimes we go in. But even the strongest swimmers among us cannot stay long in the coldest water, not without fur, feathers, or blubber to keep precious body heat from draining away. The real miracle of boats is simply their buoyancy. To be aboard one is to walk on water.

These days, more and more canoeists are deliberately getting wet. They launch sleek, synthetic, nearly indestructible craft into boulder-strewn channels where spring runoff churns and eddies around the rocks. At worst they capsize or wrap their canoes around bridge piers. At best they end their runs soaked and exhilarated.

I do not begrudge them their adrenaline rushes, the surges of pulse and energy. They are playing the way they work. But my son and I have swapped whitewater for backwater. A stone's throw from Walden Pond, we float in the sunlight. A dragonfly goes by, stitching the water repeatedly with the tip of her abdomen, sowing eggs. Newts, whirligig beetles, water boatmen are the things that hold our attention. We are students of bent stems and mirrored clouds—refractions and reflections. In this cocoon, like a hammock hung midway between water and air, we sit out the human race.

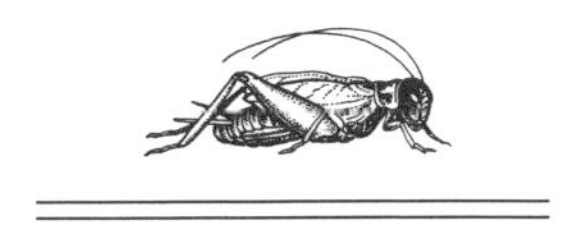

FULL POCKETS

I swim only where I can skinny-dip. I take to water of moonlit northern lakes, the deep pools of tropical streams. Here, amid water striders and fallen blossoms, I rest spread-eagle on my belly, my nose just above the surface, taking in a frog's-eye view of the world. I have no use for bathing suits, because most of them have no pockets. Without pockets, I might as well be naked.

I am by nature a collector, and depend on pockets. As a child, I was a walking natural-history museum, my slight frame improbably padded by the flora and fauna of eastern Massachusetts, plus tools for collecting, containers for storing, and food for the staff. My favorite clothes were pants advertised as having "cargo pockets." But even pockets so spacious imposed limits. Forced deaccessioning occurred again and again, if only so that pants could be washed.

Today most of the things I collect are too big for my britches. As a result, people are more likely to comment on the string of beehives, the pile of telephone poles, and the heap of horse manure in the backyard than on what makes my pants ride so low. But my pockets are no less crammed. I load them up when people aren't looking, and the only times that I am forced to make full disclosure of their contents are when I have set off the alarm at the airport.

The contents of my right front pocket alone are enough to sound the tocsin: keys for every lock I encounter—from a university gate to my barn—exact change for anything, and a Polish Underground Ballooner's Knife. (The latter I christened in the face of everybody else's Swiss Army knives.) Here I also carry a couple of horse chestnuts, a beach stone, or some other soothing talisman. Balancing the load on my left are a wallet and date book, big enough by themselves to bulge the flank.

The back pockets of my trousers are underutilized. I discovered long ago that accidentally sitting on something usually breaks or kills it. I tend to forget what is in my back pocket because I don't have enough nerves in that portion of my anatomy to retain a very clear mental picture of what I am carrying. This lack of feeling gives me real appreciation for Raymond Chandler's "She gave me a smile I could feel in my hip pocket."

Whatever the deficiencies of my posterior, I have made practical use of having to wear a coat and tie. I could do without the tie, but the coat has added dramatically to my storage capacity. Here is room to put the wadded-up tie (and yesterday's, too). Here I keep pens and pencils, notepaper and index cards, the tools of my profession, the ones a traveler of finite memory needs to hang on to a good idea. In one of my coat pockets I always try to carry a large plastic bag and a small empty envelope, in which to put objects that are too big for or might get lost in a pocket. Elsewhere, I try to carry something to read, some fruit to eat, some gum to help my ears pop, a spare shoelace, matches, and Kleenex.

But most of what I carry is not provision for the future; it is the residue of the past. I wish I were as compulsive a curator as I am a collector. The fact is that I am not always sure what I am carrying. On a typical day I can reach into one of my coat pockets and find a packet of Thai eggplant seeds, two macadamia nuts, a gastropod shell from a Costa Rican beach, a program for a Pete Seeger–Arlo Guthrie concert, two foil-wrapped chocolate mints, a large machine bolt, the packing slip for an order of gooseberry bushes, the announcement of a Cambridge Entomological Club meeting, a handful of gravel my son asked me to carry for him, four boxes of matches from different restaurants, and some unroasted coffee beans. Given a free hand, my pockets today still pick up a reasonably representative sample of my world.

There are costs to such acquisition. Chief among them is a lumpy silhouette. Women, I find, aspire to pockets, but are often afraid to use them for fear of ruining their profiles. Even men have started carrying bags in the interests of smooth hips, sacrificing convenience for style. I look the way I always have, and it doesn't bother me. If it did, I suppose I would have my clothes custom-made, remembering the English gentleman who was asked by his tailor, "About the hip pocket, Sir, pint or half-pint?"

Using my pockets to their fullest allows me to be generous in offering pens and pencils, paper, matches, the loan of a knife. And I am always willing to take on something extra, someone else's keys, makeup, a scarf, since they add relatively little to the load. But these are minor benefits compared with the real pleasures of full pockets. My favorite stories, small celebrations of people and nature, so often begin with a common object, some holdfast to which I can anchor a tale. It need not be exotic, a galvanized-metal spout used for maple sugaring will do. If I am short of inspiration, I turn out my pockets. The more they contain, the more likely I am to find the start of a story—an electric-fence insulator, a strip of Chiquita Banana labels, some heirloom beans. Do I write because I have filled my pockets? Or is it the other way around? It does not matter. So long as I do one, I will most likely do the other. I can be expected to go hand in pocket into the sunset. "Our last garment is made without pockets," says an old Italian proverb. I prefer to say that someone without pockets is someone without life.

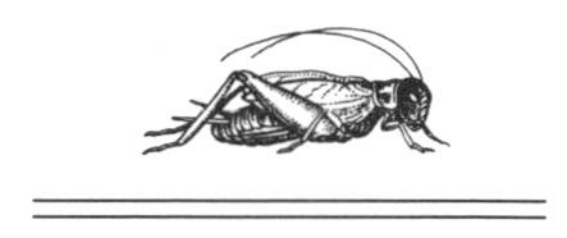

THE ONION BRAIDER

He likes braiding onions. He likes braiding tight, adding in an onion every second or third plait, putting as many of the flat yellowish-brown bulbs on each string as he can. Dumping the onions into a mesh sack and hanging the sack by its drawstring would keep them just as well, but these onions deserve to be done up fancy. They are his onions, and, as he touches them, he remembers the times he brought them water, carefully unrolling the hose so as not to drag it over the neighboring lettuce. June and July were dry, so dry the lilacs wilted, and water was as scarce as it was needed. To get the most out of what water was left, he scraped up a low dike of soil along the downhill side of the onion row so that the narrow stream of water flowing from the hose end stayed close to the bases of the young plants and soaked in.

Judging from the size of his onions, they had enough of everything—enough sunlight and enough fertilizer as well as water. And all of these provided so that individual onions never had to fight to get what they needed. Over and over during the summer, he stopped and stooped to pull out even the smallest weed from the row, careful not to knock over the delicate blue-green tubular leaves of the onions themselves. There was plenty of space for weeds to grow, for he spaced the onions a full hand's breadth apart, but he tolerated no interference from volunteer seedlings of any kind. "To grow competition onions," he liked to say, "you have to grow them without competition."

Three weeks ago, with summer cooling down and the first swamp maples beginning to color, he went down the onion row with a broom, knocking over the few remaining tops that had not fallen already. And the day before yesterday he went down the row again, pulling the onions up, laying them on top of the soil with their tops all in one direction, looking as though they had been combed out by the receding tide.

Now the tops are a mix of green and brown, brown enough to be pliable, green enough to be strong. If he waited longer, the tops would become so brittle as to be unbraidable. Even now, he weaves in a length of

baling twine, passing it from hand to hand along with the strands of onion tops. These onions don't need the reinforcement, but using the twine had become a custom of his, partly because it is easy to tie the sisal into a loop for hanging the braid up when it is done.

He likes braiding onions. Sitting cross-legged on the cooling ground surrounded by this year's harvest, he remembers a girl's hair he braided once—long, shiny brown hair, smelling warmly of shampoo. Most of the women he has known since have had short hair, the length of fashion and convenience. His wife's hair, though she leaves it uncut, has never grown long enough to braid properly, and all his children are sons.

Braiding onions has become an autumn ritual of his, something he fancies he is pretty good at. Of course, he would be the first to admit that there are probably onion braiders who are better. In England, perhaps. He has heard that in England there are people who double dig their gardens every spring and spend their lives trying to grow a parsnip as big as a man's leg or the best leek in all of Lower Bottomly on the Marsh.

In the years just after he began growing vegetables, he got caught up in a bit of this sort of competition. For several seasons he grew everything in the catalogue—dozens of different tomatoes, and nearly as many cucumbers, lettuces, and winter squash—entering the best of each kind in contests held at various county fairs in the state. Somewhere he still had a shoebox full of ribbons, proof that he had been pretty good at both growing and showing, but if the truth be known, the competition wasn't so stiff that the prizes meant much.

Since then, his gardening has become simpler. This year, he planted only four different tomatoes, two cucumbers, a single squash. He no longer tries to have peas by the Fourth of July or a second crop of lettuce in the fall. He has stopped dragging hundreds of pounds of old wooden storm windows out to the garden each spring to build glass pup tents over the tomatoes, for, as he says, "open a quart of stewed tomatoes in February and who cares whether they ripened in July or September."

The only onion he still grows is called Stuttgarter. He grows them from sets, not seeds. The sets are dime-size bulbs grown the previous year by companies that sow the seed late in the season. The long days of summer cause the plants to form bulbs before the plants have had much chance to grow. Stored over the winter, these onion sets can be replanted in the spring and will resume growing, ultimately producing a full-size onion. The Stuttgarter onion sets he buys he gets from the hardware store, two pounds of them every spring, roughly four hundred individual sets.

Onion sets, he has found, have the virtue of yielding an onion for every one planted. And they do so consistently, even in his garden's abbreviated growing season. Full-grown Stuttgarter onions are fist-sized, not as big as White Sweet Spanish onions or as colorful as Red Hamburger onions, but he has found they keep better. By hanging the braids of onions from an old hat stand on the landing of an unheated back stairway, he can count on having onions year round. What the Stuttgarter onions lack in size, they make up for in authoritativeness. A single bulb is all the pungency one wants.

He may no longer grow as many vegetables as he once did, but at least he still gardens, he reminds himself occasionally. So many of his men friends have put their gardens back to lawn. Some decided they preferred golf. Others, once they had grown everything once, moved on to new frontiers. Some gave up when they realized that unless they put food away, a vegetable garden fed the family for only ten weeks a year. Others quit when they stopped and calculated how small a dollar return they were getting for their effort.

Asked why he continues to grow vegetables, the

onion braider will only say that he likes the work. But this is because he is hesitant to admit he is awed by his ability to turn bare ground green, by the way he can move his hands and bring forth abundance. Summer after summer, he has watched the garden fill up with foliage, until by fall he can never remember quite what it looked like when the snow melted. The corn, the beans, the pumpkins are all part of the change, but none delight him as much as the onions. They are so predictable. He has never had an onion failure. He has never heard of anyone else having one. Not with onion sets. Onions grow. He has come to count on it. It is the certainty that they will grow that inspires him to put the extra time into raising them, the extra attention that brings out their best. He cannot imagine a garden without onions, just as he cannot imagine a fall that did not include an afternoon in September when he sat on the ground braiding their hair.

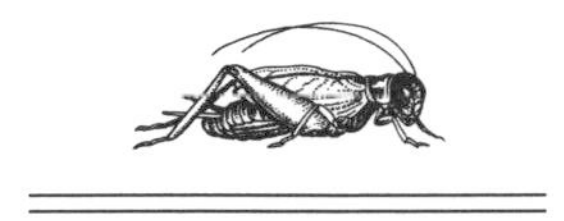

BLUE BIRDFOOD

This summer's big concern has been the disappearance of trees. Rain forests have been getting most of the press. The newsmagazines have all run pictures of loggers working in the Pacific Northwest—big saws, big trunks freshly felled. The photographs of the Amazon basin have been smoke-filled, with tongues of orange fire licking the slash and leaving behind cemeteries of charred black stumps.

There has been much heated rhetoric about imports and exports and saving the jobs of the tree cutters. About the only issue on which everyone agrees is that wood is too precious to be going up in flames. The thermometer outside the kitchen window has been reading hotter than usual, as if to record the intensity of this debate. And some people wonder whether we are feeling the heat of the tropical conflagrations themselves.

Perhaps we are. Bird census takers—the sort of bird-watchers who keep track not only of what they see but how many—have been reporting a drop in the numbers of summer songbirds. And because so many of our birds (40 percent of North American species) migrate to and from the tropics each year, it is easy to blame this decline on the torching of their winter homes. Those who have studied these birds in Central and South America report that the winter arrivals, from indigo buntings to yellow-bellied sapsuckers, not only set up winter feeding territories but return each year to the same ones. Clear-cutting, as the critics point out, consumes more than cellulose.

Of course, not everyone is convinced that it is the chop-chop of ax and machete that is drowning out the dawn chorus. Science is seldom so cut and dried. You need only put a condominium complex for purple martins on top of a tall pole and wait in vain for a full house and a mosquito-free backyard to discover that it is the nature of birds to outwit theories. What I have learned from twenty-five years on this wooded New England hillside is that making a hole in the forest can sometimes lead to an avian excess. Witness the fact that our family is doing such a brisk business in blue birdfood.

Ours has been a case study not of the effects of an encompassing deforestation on birds but of reforesta-

tion, of trees growing up while we struggle to keep them cut down. This hundred-acre farm is located halfway up one side of North Pack Monadnock, a 2,278-foot mountain in southern New Hampshire. At one time 85 percent of this entire state was cleared for human use. But the soil the early residents uncovered when they stripped away the oaks and chestnuts was—and still is—shallow, acid, and stony. It's the sort of ground that wears out spirits as well as tools, and once the deeper, sweeter, rock-free lands of the West were opened, these glacier-scraped hills were abandoned.

Among the first native shrubs to spring back up, once the plowing and grazing stopped, were blueberries: lowbush on drier ground, highbush in the swamps. For the families left behind by the westward migration, the wild harvest must have been some consolation. At least it was a change from picking stone. The old and young took to the hills every summer. Some, I have heard, could remove the fruit so fast that a steady stream of blue fell into their baskets and pails.

With the arrival of the railroad in 1874, blueberry picking became big business. The storekeepers provided the wagons, drove the pickers to the foot of the mountain in the morning, brought them back in the afternoon, and purchased and packaged the berries they had gathered. Records for 1892 show that twelve hundred crates of berries were shipped to Boston, fattening the local economy by a thousand dollars.

In 1905, a U.S. Department of Agriculture botanist from Washington, D.C., named Frederick V. Coville began summering here with his family. The next year he started his famous experiments on growing and improving blueberries, research that led to today's commercial industry. It was Dr. Coville who discovered that blueberries not only like but require soil with a pH of 4.0–5.2. And it was Dr. Coville who began to breed better blueberries. He started with a highbush berry of particularly fine flavor that he found in a brushy, mountain pasture and named 'Brooks' after the owner of the land. He crossed this with an early-bearing lowbush named 'Russell' from another nearby farm. Every muffin, every pancake made today from cultivated berries contains a little of this first blueblood.

When we moved to this farm twenty-five years ago, the entire top of North Pack Monadnock was still open land, for sheep had been pastured there until the 1938 hurricane. Low stone walls crisscrossed hundreds of acres of blueberries, and crowds of people made pilgrimages up the mountain to pick. In the years since, however, chokecherry and gray birch have been steadily seeding in. Red maples and white pines now shade even the highest blueberry bushes, and it is mostly hikers who use the trail. In another generation or two, the trees will be so tall that even the views will be gone.

On our own land, on the mountain's shoulder, we have spent much of our free time maintaining a "state of arrested succession"—a condition intermediate between hayfield and oak forest in which blueberries can thrive. Remove an adolescent white pine, let sunlight fall on a fifty-year-old blueberry bush, and it will flower and bear fruit again. Keeping the forest at bay is hot, pitchy work, but in the early years there was always the solace of pie. True, some years were more bountiful than others, depending on the harshness of the winter or whether it rained when the blueberries were in flower and kept the bumblebees from pollinating the pinkish-white bells. But lately the weather hasn't mattered. The bushes can be heavy with green fruit in June, and there will be precious little to pick come July. It isn't that we have fewer bushes. In fact, we have been planting some of Dr. Coville's better descendants. But we once put fifty or seventy-five quarts of blueberries in the freezer each summer, and now we have to scramble to find a dozen.

The problem is birds. To human tastebuds a blueberry doesn't reach full flavor until a week after it has turned completely blue. Avian palates must differ from ours, for the birds start eating blueberries almost as soon as the fruit starts to ripen. And birds are born pickers. Flocks of robins and rufous-sided towhees, catbirds and bluejays, fill our bushes. We've counted hermit thrushes, wood thrushes, brown thrashers, flickers, and cedar waxwings. Even if we don't see them feeding, we can tell where a bird has been dining by the blue droppings.

We like birds. They are, after all, the natural disseminators of blueberries, raspberries, and other small fruits. And one of the reasons we encourage vigorous thickets on our land is that they support wildlife. We will never tire of spotting bluebirds, scarlet tanagers, or northern orioles no matter what they are feeding on. Nor do we begrudge sharing the harvest with ruffed grouse, or the newly returned wild turkeys.

But birds don't know much about sharing. This spring we realized that this birdfood business of ours was headed straight for a total absence of blueberry muffins on our breakfast table. The problem is that the blueberries the birds once fed on up on the mountain are by now so shaded by trees that the birds must concentrate on the few fruiting patches left, namely ours. Given time, the birds and blueberries will no doubt come to equilibrium. But in the meantime I have more than once been tempted to reach for that premier bird-studying instrument of years past—the double-barreled shotgun. Scarecrows don't work, cannons are too noisy, and we were never willing to spray a chemical that was said to make the berries taste bad to birds. The description alone did the same for us.

The solution we have resorted to is netting, seventeen-foot-wide lengths of black polypropylene with a three-quarter-inch mesh, joined at the edges with a lacing of baling twine. Our blueberry patch now looks like some great environmental sculpture. Because the netting catches on berries the way shirt buttons do on a loose-knit sweater, and because birds can still stand on it and peck through, we have it supported by a grove of six-foot-high ash poles linked by wires at the top. Log sections seal the edges of the net to the ground.

We can't afford to cover all our blueberries, even if we wanted to. The netting is strung over the descendants of Dr. Coville. We are leaving their wild antecedents to their traditional pickers, the birds. We will have to take the netting down before winter, and put it up again next summer to extend its life. But judging from the way we feel now, no one will complain about the chore. Not when inside the netting the blueberries can ripen in peace. They can stay there until they fall off of their own overripeness if we choose.

What has been saved by this protective envelope, we have discovered, has been more than our breakfast. Yes, we are looking forward to bigger and bigger harvests as the bushes under the netting reach full size. But more important, we find we have single-handedly turned down the heat on the issue of whether the world belongs to us or the birds. Biologists, economists, loggers, and bird-watchers will continue to debate the consequences of cutting down trees. And we, not without sawdust on our hands, will be joining in. But for a moment, in a long, hot summer of controversy, we intend to sit back and enjoy this cease-fire.

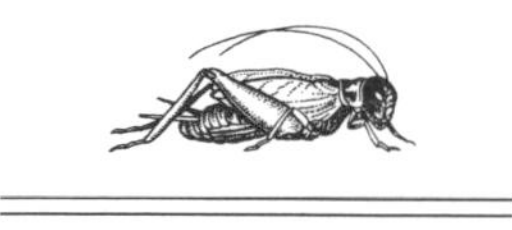

HAND TO TOOTH WORK

I am often mistaken for someone who does not own a comb. This is not true. I have a perfectly serviceable comb on my person at all times. If I arrive at my office looking windblown, it is only because I have forgotten to use it. In response to some recent comments to the effect that I am, if not sharply dressed, at least better kempt, I say that I have been using my wife's comb. It is a new one, and she keeps it on top of her bureau where I can't miss it when I get up in the morning. I feel not even the slightest twinge of trespass as I run her comb through my cowlicks. Indeed, I feel positively entitled to use it. I am, you see, its maker.

My own comb has unbreakable teeth and is made of red plastic, the sort of red that you can spot at a hundred yards should you drop it accidentally. Although I have carried this comb around in my hip pocket for years, I don't know much about its history. About all that I can say is that it originated in an oil field somewhere, but whether in Texas or Saudi Arabia or the North Sea is anybody's guess.

On the other hand, I know exactly where Elisabeth's comb came from. It came from two trees—a black cherry and a red maple—in a grove we cut for firewood two years ago just downstream from our sugarhouse. We sawed the trees down, cut them into four-foot lengths, split them with a sledgehammer and wedges, and piled them bark-side up to dry. The next fall, we cut the seasoned wood to stove length and restacked it in the shed.

At the time, in the flurry of sawdust and wood chips that accompanies the rush to get the winter's wood in, none of us saw the comb. But winter is my season for intellectual chores, and spring is many paragraphs off. When you alternate putting words on paper and wood in the stove, you notice things like the differences between one piece of tree and the next. Most of the firewood is knotty, twisted, and ragged, but now and then you come across a piece that has split so cleanly that the faces are as smooth as a board. It does not take much to see that the latter is an example of the fine hardwood about which cabinetmakers wax eloquent.

The pieces of Elisabeth's comb were a handspan away from incineration when I saw them in a different light and unceremoniously dumped them down the cellar stairs instead. Some days later, frustrated by a recalcitrant phrase, I stopped typing and went down cellar to see about turning cordwood into lumber. For my birthday a few years ago, Elisabeth, knowing my tastes, gave me a froe. A froe is an L-shaped tool, about the size of a rafter square, one arm of which is a stout wooden handle and the other a heavy iron blade sharpened on the outside edge. The demand for froes today is about the same as for ice saws, but if you want a tool that gives you a controlled split, you want a froe. Once the blade has been driven into the end grain of a piece of wood, the handle gives you leverage to continue the split. Froes have many uses, and many common names, one of which is board ax.

To start the split, you are supposed to beat the froe with a froe club, a hunk of hickory trunk and root that looks like something you might use to slay a woolly mammoth. I don't have a froe club, but I do have lots of firewood. When the piece with which I have been beating the froe blade begins to splinter, I reach for another. Even with a froe, the wood doesn't always split the way you intend. I have learned to split off rough boards that are thicker than what I ultimately want.

Next, I clamp each of the rough boards in a vise, and plane it until it is smooth, square, and uniformly thick. You can use a block plane to do this if that is the only plane you have, but the longer the plane, the easier the job. Back before woodworkers had the motor-driven machine called a jointer, they used a jointer plane with a twenty-four-inch sole to level the surfaces of large pieces of wood.

Planing is like ice skating. It looks a lot easier to do than it is, and perfection requires that even an expert concentrate. But the rewards are immediate. As the wood curls up from the plane's iron, the board in the tree begins to appear. After only a few strokes, you can see the wood's real character. The pink heartwood of the cherry begins to shine. The maple reveals those minute imperfections that make it bird's-eye or curly.

About this point, I begin to lose steam. A combination of winter-softened muscles and the knowledge that none of this is getting any writing done makes me put the partially planed pieces of wood away under the workbench. This is just as well, for the wood that is dry enough to burn still has a lot of moisture in it from a woodworker's point of view. Cabinetmakers must be nearly as farsighted as foresters. An inch a year was the Shaker rule of thumb for the time it takes lumber of varying thicknesses to dry.

When the pieces of maple and cherry went under the bench, I hadn't the faintest idea what I was going to use the wood for. I wish that I could look at a plain board and see some future object lying within, as

Michelangelo looked at a marble block. Oh, I see things, all right, lots of different things all superimposed. Unscrambling the various outlines is like trying to read a piece of carbon paper that has been used again and again.

Nevertheless, I like making things from scratch. The nicest thing you could say about a cake when I was growing up was that it had been made from cold water and air.* I would always rather start with a piece of log than with a bought board. In this I am not alone. Craftwork has a persistent allure for those of us in coat-and-tie professions. The attractiveness of making woolen cloth from sheep fleeces or furniture from tree branches is, I think, a reaction to the increasing specialization of our work. We jump at any opportunity to shepherd something through from start to finish.

Like many, I dream of making a living with my hands rather than my mind. But until then, I remain a part-time woodworker. The overriding demands of my real work force me to be selective. Like a weekend cook, I page through books and magazines looking for something that matches the time, ingredients, and skills I have.

*This is the way I heard it as a child, but my parents, a pair of organic chemists, were actually saying "coal, water, and air," the cornerstones of synthetic chemistry.

The idea for a wooden comb came from an article in *Fine Woodworking* about a man who gave up white-collar work a decade ago to make wooden combs full-time. Most such combs are made from a single piece of wood with the grain running in the direction of the teeth. Sooner or later, each of these combs breaks in half across its spine where the grain is weak. This combwright's innovation was to surround the wooden teeth with a second piece of wood in which the grain runs perpendicular. When two contrasting woods are used, the resulting comb is both strong and beautiful.

Although this combwright had invented ways to fashion his combs fast enough to make a living at it, a note appended to the article said that it was possible to make just one comb in about four hours using ordinary workshop tools. This was the signal for me to go downstairs and set about marrying my pieces of maple and cherry into a present for Elisabeth.

I won't go into all the techniques I used to make my comb. Suffice it to say that I was only half done at the four-hour mark. I used a lot of tools, some more efficiently than others, some probably for things for which they were never intended. I used five-minute epoxy to glue the maple tooth blank into the cherry handle. I used a table saw with a plywood blade to rough out the fourteen teeth. I used a huge amount of elbow grease filing and sanding and polishing. I had a marvelous time.

My enjoyment was to be expected, given that at the time I was happily avoiding my desk. What is remarkable is that the finished comb should continue to delight me so. It has a sort of magic, this phoenix risen from almost ashes. In this scant handful of silky hardwood, I can see the standing trees from which the comb came as clearly as I see the woman for whom it was made. In the graceful curves, the polished surfaces, and precise teeth, I find proof that paragraphs are not the only things worthy of attention. This wooden comb is the best morning pick-me-up I know, a fresh reminder not to take my life as a wordwright too seriously. I need only run this comb through my hair once, and I no longer have to be at the office on time. Or if I do, at least my hair will be combed.

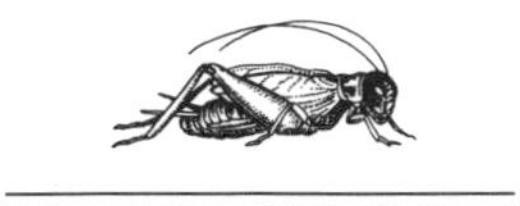

DEADSTOCK

The animals I husband all have six legs. They aren't freaks, like those creatures with extra limbs you pay fifty cents to see preserved in dusty jars of formaldehyde at county fairs. They are ordinary honeybees. For a lapsed shepherd such as myself with a short attention span and a host of other obligations, bees are the perfect flock. I cannot imagine another domestic animal that I would rather watch over.

Pigs have their champions. But no one has ever had to slop a honeybee. Nor do bees need protection from marauding dogs the way sheep do. Bees don't tip over and dirty their water like rabbits. They don't stink like goats. And compared with a flock of chickens, they are silent.

No honeybee ever broke down a fence—or needed one in the first place. Bees forage miles away from the hive in all directions, freely trespassing on neighboring land in search of food and water. (Being virtually indistinguishable from one another, they have built-in alibis.) Unusually provident beasts, honeybees are not content with procuring their daily bread, but go on gathering against a rainy day or a long winter.

Unlike horses or cows, bees muck out their own stalls. They patch leaks in the hive roof and holes in the walls. They heat the interior in winter and air-condition it in summer. If this weren't enough, bees serve as their own midwives, nurses, and watchdogs.

What I like best about honeybees, however, isn't the way they live, but rather the way they die. Anyone who raises animals must face the responsibilities of death sooner or later, if only because humans live longer. People regularly celebrate their hundredth birthday, but the maximum recorded life span for a horse is forty-six. The record for cows is thirty, the same, surprisingly, as it is for chickens. A pig once lived to be twenty-seven, but sheep and goats have only reached twenty and eighteen years respectively.

The only sure way not to outlive your livestock is either to hold off getting animals until you retire, or else raise large tortoises. Marion's Tortoise was already adult when the French explorer Marion de Fresne moved it from one of the islands in the Indian

Ocean to the island of Mauritius in 1766. It died in 1918, having reached an age of at least 152. Another tortoise, named Tu'imalila, reputedly presented to the Queen of Tonga by Captain James Cook in 1772, lived until 1960, a dozen years shy of two centuries.

Extreme longevity offers an escape from death duties, but this isn't of much use if you are raising animals to be eaten. An animal being raised for its meat is almost invariably slaughtered in the prime of its life. Most of us who aren't vegetarians have learned to accept this death, however, as the price of our own carnivory. The fact that the death is deliberate and anticipated from the start makes it somehow easier to face. Or as E. B. White wrote from his farm on the Maine coast: "The scheme of buying a spring pig in blossomtime, feeding it through the summer and fall, and butchering it when the cold weather arrives, is a familiar scheme. . . . The murder, being premeditated, is in the first degree but is quick and skillful, and the smoked bacon and ham provide a ceremonial ending whose fitness is seldom questioned."

Much harder to deal with, I have found, is an unexpected death. Even when an animal is raised for something other than its meat—for milk, eggs, wool, feathers, or simply the strength of its muscles—it can fail to live out its natural life span. A cow slips through a barbed-wire fence bordering a highway, walks down the yellow line, and is hit by a car. A horse and her colt are electrocuted by the ground shock radiating outward when the tree under which they are standing is struck by lightning. A lamb is disemboweled by the two collies and the German shepherd that live down the road. A pig contracts pneumonia and is dead a few hours after you first hear it sneeze. Sometimes, there is no apparent cause of death. Livestock simply turns into deadstock.

What do you do with a dead cow? An animal that has died of unknown causes is presumed unfit for human consumption, and even one killed by accident is often unsalvageable. Disposing of the carcass is all the more critical if disease is involved. My handbook of animal husbandry and veterinary care says the body should be buried six feet down. Where I live, bedrock is shallower than that, and four months of the year the ground is frozen solid. Farmers who lack a backhoe or a blaster's license, and who can't persuade someone to haul the body away, may have to burn it. Supposedly, if you raise the carcass up on three or four logs, cover it with straw and kindling, and drench it in motor oil or kerosene, a cow will burn. I don't want to find out if this is true.

And yet in the midst of a crisis, which sudden death always is, we are all capable of devising solutions—tragicomic as some may be. There is the story about the New York City resident who, when his large dog died, wedged the body into a suitcase to take it somewhere, and had the suitcase stolen from the sidewalk while he was hailing a cab. Our own beloved guinea pig, the last of a procession of family animals going all the way back to chickens, had to be temporarily laid out atop the chocolate fudge whirl, between the orange juice and the English muffins in the kitchen freezer, until ground could be broken for a more dignified burial.

All of this talk of dead cows and other animals is a way of emphasizing how fond I am of honeybees. They are not particularly long-lived, though the queen may survive four or five years. Quite the opposite is the case for most occupants of the hive. During the summer, the female worker bees live only five to seven weeks (excluding the three weeks it takes to go from egg to larva to pupa to adult bee), the male drones only a week longer. Workers who winter with the queen may live four to six months, in part because they aren't working as hard and in part because their physiology is different from that of summer workers, but once warm weather returns they, too, pass on.

To calculate how many bees die each year in a hive, you need only look at how many are born. Honeybee queens are capable of laying fifteen hundred eggs a day. In a strong colony, perhaps two hundred thousand new bees are produced every year.

Since most beekeepers discourage swarming, for every bee born another bee must die. I currently have three hives between the crabapple tree and the back fence, a modest endeavor as beekeeping goes. And yet I have more than half a million bees dying every year.

Half a million honeybees weigh in aggregate almost as much as I do. But no one has ever had to dig a grave for a bee. Ninety percent of them die individually, unsung and unnoticed in foreign fields. Foraging is dangerous business. Ordinarily, workers do not even begin until their twenty-first day. The preceding three weeks are spent in the relative safety of the hive, feeding the young, cleaning house, constructing honeycomb. But once they start leaving the hive, bees get caught in spiderwebs and rainstorms. They are eaten by birds and poisoned by pesticides. Many of them simply wear out. Every honeybee seems to be warranteed for a certain number of miles of travel. The sooner a bee covers this distance, the sooner it dies. Now and then, you find a solitary worker on the ground, her wings frayed; still alive, but incapable of flying back to the hive. She is just one of about a thousand from her colony who perish each day.

Though most bees die away from the hive, some fifty to one hundred a day succumb inside the hive. If the bodies were allowed to remain, the accumulation, amounting to about a quart a month, would soon become unhygienic. But here again, honeybees separate themselves from sheep and goats. When a worker collapses in one of the dark, honeyed halls of the hive's interior, the body is swiftly removed from the hive, usually in less than an hour. Another bee grabs the body in her mandibles, drags it outside, and sometimes flies as much as a hundred yards away before jettisoning the corpse.

What I know of this comes from an article titled "The Honey Bee Way of Death: Necrophoric Behavior in *Apis mellifera* Colonies" by Kirk Visscher, a graduate student at Cornell University. The author, an even greater connoisseur of dead bees than I, has discovered that within each hive there are workers specializing as undertakers. These bees, who make up 1 to 2 percent of the population, apparently identify their deceased hivemates by smell. Or at least Visscher found that dead bees coated with paraffin or purged with chemical solvent were removed from the hive more slowly than the untreated dead. The bees who serve as undertakers do so for several days near the end of their period of housekeeping duties, at the point when they have taken their first orientation flight and are about to become foragers.

Much remains to be learned about how bees die, but I will let others tease out the details. I will content myself with spending idle hours on warm spring afternoons leaning against the sunny side of one of my hives, watching the entrance around the corner. On days like these, a funnel of bees extends skyward from the narrow ledge of the bottom board on which the bees land and take off. Most of the incoming workers are loaded with nectar. Others are carrying paired balls of yellow, orange, or red pollen on their hind legs. As they land and scramble past their sisters guarding the entrance, they pass a stream of outgoing workers, each of whom lifts off and climbs swiftly skyward.

The longer I watch, the more I am drawn into this golden vortex, caught up in the microcosm of the hive. "Death eats up all things, both the young lamb and the old sheep," wrote Cervantes. Honeybees are no different. But in their great numbers and brief life spans, it is easier to see that death and life are part and parcel. There is no need to fear the one where there is a constant reminder of the other. The bees safely returning with provisions stand for life. The bees departing on their perilous flights stand for death. In the compressed time scale of a honeybee colony, life matches death. All is balanced. And as I look up in the air to where the funnel spreads out and the individual bees are nothing but tiny silhouettes, I can no longer tell which bees are coming and which are going. Life and death become indistinguishable against the blue sky.

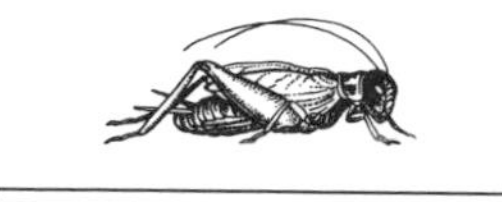

WALKING BOUNDARIES

It is time to walk the boundary of our land again. The route is neither long nor hard to follow, just over two miles of old stone wall that neatly divides our forest from our neighbor's. The girth of the oaks and hemlocks that crowd against the wall's length makes one think that the wall must originally have been built through the woods. But when its boulders were lined up, the land on both sides was open fields and pastures as far as the eye could see. Sun shone down on the men and oxen when the boulders were rolled onto wooden stone boats, dragged here, and levered into place.

At least once a year, I like to circumambulate this property, letting the lichen-covered wall be my guide. I want to see what is new—whether beavers have ventured upstream, whether the pitcher plants are blooming, whether blue herons are back. I try to check up on a cohort of hemlock seedlings growing out of a rotten log, on a pine tree that ants and pileated woodpeckers have been gradually demolishing. Beyond the things that I half-expect to find are always a few surprises—a weasel popping out from under the wall, a beech struck by lightning. Long before I have completed the circuit, I am glad I took the time.

The time, unfortunately, is hard to find. Like a woodchuck, I seldom venture far from one of the entrances to my residence. A hawk's-eye view of the farm would show the well-worn paths radiating from house and barn—paths to the vegetable garden, the compost heap, raspberry patch; paths to the woodpile, the orchard, vineyard, and sugarhouse. When our family decided to cultivate this bony land once more we little dreamed how it would shrink our horizons.

Even from my myopic perspective, however, I am beginning to see that I must stop thinking of boundary walking as an afternoon's diversion, something that can be put off in favor of canning corn or patching the roof. This land is currently in danger of undergoing a change far greater than that wrought by the 1938 hurricane, which blew down so many trees—greater even than that wrought by the builders of the stone walls who, before that, cut down them all. A wave of asphalt, spawned by a bulging city, is rolling up the valley beneath us. The asphalt carries factory outlets, fast-food franchises, bottling plants, and automobile dealerships. Dairy barns are being crushed, and the river-bottom soil, where corn once ripened on hot August nights, is being bulldozed into huge piles. Soon it will be scattered around the little houses that the surge of asphalt is splashing up on the surrounding ridges.

To see property only a few miles away being transformed with such speed makes me anxious to walk the boundaries of my own land again. Not that a shopping mall could have appeared mushroomlike in the back forty overnight. The worst that is likely to have happened is that an overeager logger has cut a few trees on this side of the wall, or that hunters have dumped the remains of their lunch. But as this wave of asphalt washes by, it will force all the landowners in the vicinity to make choices.

I will opt for conservation, but if this property becomes an island in a sea of asphalt, it will preserve little. There must be many such tracts, in order to save a significant part of the region's plants and animals. All of them will have to be protected against short-sighted exploitation. Some proponents of development will claim that "wildlife refuge," "open space," and "endangered species" are nothing more than empty words, piled up to block progress. The words themselves are empty: They have meaning and usefulness only in proportion to the amount of detail that backs them up. We must be prepared to describe what it is about wildness that we want to preserve. It is easy to forget the details, to forget about duck nests and walking ferns. The best way to remember, the best way to restock with a fresh supply of reasons, is to walk a boundary again.

All of us have boundaries that we can walk, whether or not we have title to land. Boundaries are, by definition, those places that we don't get to visit very often. They are nevertheless places that are important to us, whether they be sources of nourishment or inspiration—a wheat field or a volcano. Our ever-increasing human population needs ever more resources to support it. Demands for energy, food, raw materials, and recreation are scarring the most remote deserts and beaches, the vastest forests, the deepest lakes. We must take the time to go and see what is happening, to find out what is new, and while we are there take note of what it is that we might lose.

It is easy to get caught up in daily life, whether it is going to the Laundromat or weeding strawberries. It is easy to say that change cannot be stopped, that the land has been through changes before, and survived.

Indeed, the land has changed and survived. The stone walls ringing my woods are proof of that. But the current changes promise to have more far-reaching consequences than any that have occurred before. Ground water is being contaminated, soil is eroding, more and more species are going extinct. Change cannot be halted, but it can be directed. A greater understanding of our boundaries will give us some of the power and the wisdom we need.

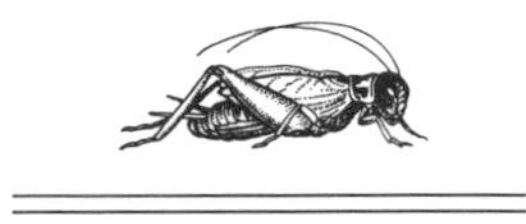

HARNESSING THE BOGS

The scientific name for black crowberry—*Empetrum nigrum*—comes from the Greek words meaning "upon a rock," a reference to its taste for high mountain summits. This alpine evergreen, with its short, needlelike leaves and round, black, edible berries, belongs among boulders and scree, not sprawled in a coastal bog alongside pitcher plants and cotton grass. It is not the bog's closeness to sea level that makes the crowberry, this plant from another place, seem so strange. Nor the air of herring and tidal flats. It is the footing. Nothing could be less rocklike, nothing so *terra infirma,* as the surface of a bog. Stop walking, whether to watch for moose or to windmill at the cloud of mosquitoes about your head, and you can sink in over your boot tops.

For all its flatness, the Acadian Peninsula has much in common with mountain peaks. Jutting out into the Gulf of St. Lawrence, this corner of Canada is bathed regularly and wrapped in thick fogs. This superabundance of moisture makes plants such as the black crowberry and the amber cloudberry feel right at home. And it is this moisture that sustains the sphagnum mosses, species of green, brown, and burgundy red that cover great expanses of ground in a thick, wet blanket.

Live sphagnum can hold as much as twenty-five times its own weight in water. Even in death, it is preserved by moisture. Under this bog's living mantle, the waterlogged ground contains too little oxygen for microbes to cause much decomposition, and so the sodden remains accumulate in a dank, dark layer called peat. The buildup is slow—as little as a fiftieth of an inch a year—but after some ten thousand years of arrested decay, the peat in places measures twenty to thirty feet thick.

Standing on solid ground at the edge of this bog and looking across to the other side, I can see only the tops of the white cedars, black spruces, and tamaracks. A dome of peat higher than the level of the surrounding ground blocks the view. Elsewhere, in drier climates, bogs may become overgrown by forest. Here it is the other way around. This raised bog is spreading, engulfing shrubs and trees at the rim, slathering their stems and stumps with a fresh layer of peat, swallowing them whole century by century.

Except for the sphagnum itself, the vegetation

atop such a bog is sparse. For despite the rainfall, this is a physiological desert. Nutrients are in chronic short supply. Pitcher plants and sundews live by hijacking the fertilizer flown in by insect bodies. The water permeating the peat is too acid for most roots to absorb. The sheep laurel, leatherleaf, Labrador tea, and other heaths that form thickets atop the sphagnum all have small, tough, or woolly leaves that cut down on moisture loss. The few scattered conifers are weather-beaten dwarfs.

But if the pickings are poor on the surface of bogs, the depths contain riches. The early colonists who drove their claim stakes at the edges of the woods, believing the quagmire beyond to be not worth their attention, could not have foreseen the time when peat land would be more valuable than dry ground. Then they could not have foreseen that gasoline engines would replace horses.

When people stopped traveling in buggies and carriages, when ice and milk stopped being delivered in wagons, when the doors on backyard stalls and livery stables closed for good, more than the horses disappeared. So did their manure—fresh manure for hotbeds, composted manure for potting soil, manure for asparagus and rhubarb. Both amateur and market gardeners had grown to depend on horse manure to improve their soil, and when it was gone, they turned to peat.

Today, a bog being harvested for horticulture has a smooth and springy surface, a great flat expanse of reddish-brown cut only by narrow, dark drainage ditches every twenty yards, which converge at the horizon. Tractors dragging harrows fluff the surface in a cloud of peat dust. Whenever the weather has been dry for several consecutive days—only a score of times between May and September—giant stainless-steel drums with three-foot-wide nozzles come onto the fields and vacuum up the driest particles of peat moss. These are dumped into stockpiles that from a distance look like the pyramid of Cheops on a brown Sahara.

What the wind doesn't lift and blow off into the surrounding forest is eventually trucked into a packing plant at the bog's edge. There a dozen cubic feet of peat moss are compressed into six-cubic-foot bales, destined for the greenhouses, nurseries, golf courses, mushroom farms, and home gardens of the United States. As heavy as these bales may seem, they contain less than half of the 90 to 95 percent moisture naturally present in fresh peat.

Even in the driest summers, harvesters can strip only an inch or two from the surface of a bog per year. The peat that requires months to gather up, however, may have taken more than a century to lay down. In all but glacial time, peat is a nonrenewable commodity. Harvesting sun-dried bales from peat fields sounds like making hay, but it has much more in common with mining very young coal.

We have only begun to dip into the vast peat reserves of North America. The United States alone has an estimated 75 million acres of peat land. Canada has 425 million. The peat, of course, can be used for more than horticulture. Pliny the Elder, writing two thousand years ago, described a Germanic tribe that gathered soil by hand, dried it, and burned it to cook food. Today, the Russians, Irish, and Finns all use peat to generate electricity. In 1989, the first peat-fired power plant in North America commenced burning a mixture of peat and wood chips on the Denbo Heath in Washington County, Maine.

We have only begun, but already the end is in sight. We have faced such riches many times before. What became of the deep alluvial topsoils and the redwood forests? As we begin to harness peat land in earnest, let us not wait to save a few scraps at the end. Better we should divide up the turf now.

Sooner rather than later, we need to affirm that some bogs will never be asked to bear our weight. Millennia in the making, peat bogs offer a rare glimpse of eternity in this swift-changing landscape. Someday I want a great-grandchild to stand in my footsteps, ankle-deep in dark bog water. And I want her to find that the place hasn't changed an inch.

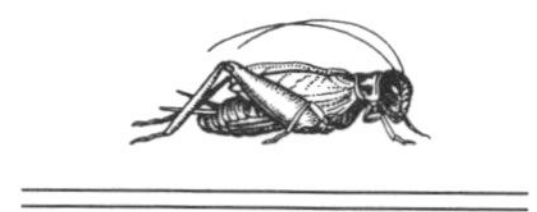

JUST WATER

My grandfather left me his taste for horseradish, for buttermilk, and for raw oysters. These fit me well. What I have only recently grown into is his taste for water. His first stop when he came to visit was always the kitchen sink, where he would pour himself a glass of water—plain water in a plain glass—drink it down, and tell us again how good it was. We knew it came from a distinguished source, the Quabbin Reservoir in central Massachusetts. We had walked atop the aqueduct through which the water flows to Boston. But none of us could see any reason to be especially excited about the liquid itself. It was just there, the stuff you drank when there wasn't any more milk.

Older now, and better traveled, I find myself standing in line, clutching my passport, waiting to be let back into the United States. In this cavernous waiting room, this legal no-man's-land, the mood is subdued. All of us, whatever the colors of our passports, are forced to reflect for a moment on the meaning of liberty. For me it has become first and foremost the freedom to drink the water. Though water may be tasteless, colorless, and odorless, in the words of Antoine de Saint-Exupéry, it is life itself.

I haven't been gone long, or far. But I am world-weary of Coca-Cola, and I have brushed my teeth with spit. The road has converted me to the one true drink: water, pure water, water that is free of foul smells or bad taste, free of algae, bacteria, protozoans, or other microscopic pestilence. I understand now with perfect clarity my grandfather's pilgrimage to our kitchen sink.

A few weeks hence, when I have drunk my fill, I may be tempted to take tap water for granted again. But the freedom to drink from the faucet must be as vigilantly guarded as any other. Attacks are reported almost daily. Sometimes the news is of a single disaster, a tractor-trailer load of industrial chemicals spilling into a reservoir. But more often the encroachment is gradual, insidious. First the water begins to taste funny, but the health department assures the residents that it is safe to drink. Then people begin to get sick. Investigations are begun, fingers are pointed, blame is assigned and denied. The water goes on tasting bad, and people are afraid to drink it.

Water purity is terribly fragile. Like minnows in a lake, contaminants spread quickly in the water, making it all but impossible to retrieve them once they are released. Entire well fields have had to be abandoned when they were salted by the runoff from nearby highways. The chemicals oozing into the earth from waste dumps can poison whole aquifers. In this country, just as in every other one, you have to watch what you drink.

This is a somewhat sobering thought for a citizen who is waiting to come home to his first real drink. So too is the realization that it is not enough to worry about the water flowing from our faucets. We must also think about the water above and below us. Every species on earth, not just our own, depends on water. To allow it to become polluted simply because we are not at the moment thirsty is both shortsighted and selfish. Unless we are more careful, hikers will contaminate every mountain stream; our city effluent will foul every estuary. Fortunately the task is as beautiful as it is important. Inspiration lies everywhere—in dewdrops and misty lakes, in marshes and tide pools.

It is my turn. I have come to the head of the line. I hold out my passport with its deep blue binding. The silver eagle ripples. The immigration officer smiles. I smile too and step forward, toward liberty and water for all.

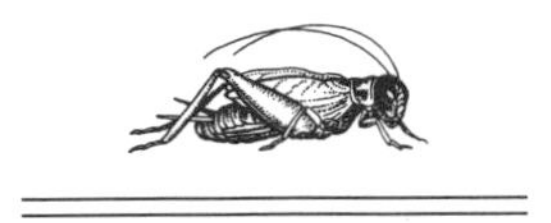

ROOM FOR BATS

Creation keeps its own hours. Spirits alight and take flight again in dark, out-of the way places. On a moist mosquito-filled night in an abandoned cacao grove, a mist net tied to a pair of poles hangs across the trail between clumps of banana leaves. A couple dozen bats, either not sensing this gossamer web or mistaking its strength, are thoroughly enmeshed in the fine black nylon filaments.

Slicing the thick darkness with our headlamps, the two of us work swiftly to untangle, weigh, sex, band, and release each animal. The last to be freed, however, is motionless. Even wrapped in my left fist for warmth, it lies still, its heart beating but slowly against my palm. A medicine dropper of sugar water presses against the tiny, leaf-nosed face. It fills with a flash of pink tongue and seconds later the heart accelerates, the legs scramble for purchase. I raise my hand. The bat is airborne.

With some eight hundred and fifty species spread over every continent except Antarctica, bats are second only to rodents in diversity and probably first among mammals in abundance. But though the ancient Egyptians thought that bats protected homes against disease-carrying demons, and Chinese mothers once sewed bat-shaped jade buttons on their babies' caps to assure long life, in our own culture these gregarious animals are almost universally shunned.

During an ordinary summer vacation in a cabin by a lake, an errant bat sweeps through the doorway of a bedroom at night. Its sighting galvanizes shrieks of alarm and a flurry of tennis rackets and beach towels. Only when the bat has been trapped beneath a hat and transported safely outdoors does the commotion end. And even then there is murmured talk between the pillows of vampires and rabies, those ghosts that lurk just beyond the porch light's yellow glare.

All of us would be better disposed toward bats were we to know them well. But we keep such different schedules. We awake after bats retire, and while we do go abroad after dusk we keep mostly to lighted ways. Bats, on the other hand, are creatures of darkness, deftly navigating the night not by eyesight but by echolocation—those pulses of high-frequency sound from open mouths that rebound to ears and enable bats to sense the precise position of stationary walls and trees and the moving targets of their insect prey. Mosquitoes, biting midges, beetles, and moths are the favored fare of temperate bats, and a single animal may eat half its weight in insects before folding its membranous wings and hanging up for the day. But it almost always does so unseen, unheard, and unsung.

Our only real cause for contact is the fondness of bats for human architecture. Loose wooden siding and unused chimneys mimic the natural crevices and hollows in which bats like to spend their days. On any house with shutters, it is a safe bet that behind one of them hangs a bat. But by far the most celebrated sites are attics. The high temperatures under an asphalt roof offer the perfect conditions for raising young, and whole colonies of bats will take up summer residence, scores of them clustered on the undersides of rafters.

Most likely these are either big brown bats (*Eptesicus fuscus*) or little brown bats (*Myotis lucifugus*). In the fall, both move to winter hibernation sites. Little brown bats go the farthest, flying a couple of hundred miles to a cave or abandoned mineshaft that remains humid and above freezing. But come spring, the females of both species return faithfully to the roosts of their birth, where they deliver and suckle their young. The males are left to lodge at large.

These summer nursery colonies of mothers and their young may contain as few as twenty-five big brown bats or as many as fifteen hundred little brown ones. But even the largest colonies may still go unnoticed by the other occupants of the house. The bats slip in and out through an unscreened gable vent or a crack in the eave. Only a faint rustling or squeaking at dusk, or on hot summer days, betrays their presence.

Then one day someone with a flashlight, hunting for some treasure long hidden in the hot darkness, spots the tiny bundles of fur, feet, and elbows. Or perhaps a windstorm tears loose a shingle and rainwater colored by its passage through the dried guano, the droppings, on the attic floorboards stains the plaster ceiling of the room beneath. Although the bats may have been there for years, seniority counts for nought. Experts are summoned and charged with eviction.

The best wait until the young are grown, construct a one-way door over the bats' entrance to the attic, and seal up all possible openings when the bats are gone. This does the job, but the dispossessed bats are left to find another place nearby to spend their days.

If we are even half-serious about having wildlife close at hand, we should be making room for bats. Before there were attics, they roosted in hollow beeches and oaks, trees with cavities a foot or more in diameter. But these have nearly all been cut down. Their loss affected lots of wildlife. Birdhouses—from wood-duck boxes on down—were early conservationists' response to a shortage of naturally hollow trees. Just recently, friends of bats have begun to build the same. Small wooden boxes open at the bottom and containing a series of narrow, rough-sided vertical chambers are being nailed up on walls and trunks where they will be warmed by the morning sun. They are modeled after bat houses in Europe, where they have been used for years. Big nursery colonies require much larger boxes, and the Missouri Conservation Department has been distributing plans for an A-frame bat house with cupola that is nearly eight feet long. When constructed next to existing colonies that are subsequently evicted, these seem to be well received.

The sawing and hammering is all admirable, but why disturb the bats in the first place and then worry about making amends? The path of least resistance is to turn out the light and open the bedroom window, to spread a dropcloth beneath the colony in the attic, to leave standing the red maple tree that is hollow and leaning across the lawn. Yes, people will wrinkle their noses when you speak of your bats. Tell them that the incidence of rabies in most colonial species is less than half of one percent, and that you don't propose to play with bats, in sickness or health, or make a practice of inhaling guano. Butterflies and birds fly fine, you can say, but I have seen the light and it is dark. Bat conservation has come home to roost.

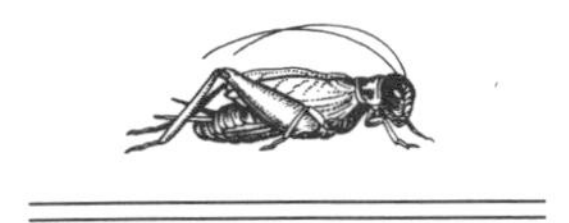

RECYCLING

Listen to the voice of a tropical forest. Not to the raucous duet of parrots, or the buzz of hovering flies, or even the love-struck chorus of frogs. Listen instead to the harmony of the falling leaves, twigs, bits of bark and fruit, the periodic crash of a large limb. Some of what falls has been intentionally shed by plants, some has been spilled by animals feeding in the canopy, some is the result of catastrophe—a branch so overloaded with ferns, moss, bromeliads, and other epiphytes that it collapses under the weight of excess vegetation.

From the sound of the forest, the ground beneath it should be deep in debris. Yet despite the steady rain of organic matter, little ever accumulates. No spongy layer of humus-rich topsoil cushions the visitor's step. To a traveler familiar with northern woodlands, with groves of oaks and aspen and pines, the soil underfoot is inexplicably bare.

What happens to all the litter? It gets reused. In the temperate zone, fallen organic matter can lie intact on the ground for years. In the tropics, it seldom lasts more than a few weeks or months. Termites, fungi, and bacteria, working in the comfort of heat and humidity, swarm over any fallen leaf or twig and set about dismantling it. Soon, thanks to the partnership of fungi and the roots of tropical vegetation, the essential nutrients that might otherwise have been washed away by rain are safely back in the forest's living tissues.

It is easy to look at the luxuriance of a tropical forest—the palms and lianas, the flowering tree trunks and plank buttresses, leaves so dense they blot out the sun—and assume that the soil beneath it all must be especially rich. But the appearance is deceiving. The original fertility of many tropical soils is long gone, stripped away by millions of years of slowly percolating rainwater. That the land is nevertheless occupied by the grandest and most complex community on earth is a monument to what can be done with strict adherence to the principle of recycling.

Those of us who reside in temperate latitudes, who are more familiar with fireweed than flame-of-the-forest, are only just beginning to know the tropics. Having been born into a land of plenty, we are still beginners at the fine art of conservation. We have been plowing through the legacy of the last Ice Age, young soil that still retains most of its original fertility. Our civilization depends on an annual harvest of grains and grasses. Yet we have been so careless with the soil's fertility that much of it has been lost. Even today, in towns and cities, we rake up the leaves that fall from the trees and bury them in plastic bags. When the nutrients from these graves surface again, they come back so contaminated they haunt us.

Our way of life is threatening our survival, and the survival of what little temperate wilderness remains. Because the next Ice Age is too far off to rescue us in time, we ought to start economizing. But making do with less is such a bleak prospect that few of us are eager to begin. That is what makes the tropics so attractive, not just to first-time visitors, but even more to those who have spent time there. Here is an example of recycling that makes our own fledgling efforts seem pitiful. Here is a fantastically complicated, enormously sophisticated world constructed entirely from used components. Even the most accomplished tropical biologists consider themselves neophytes. But if an overriding wisdom is emerging from our preliminary studies in tropical forests, it is that salvage and salvation have the same roots.

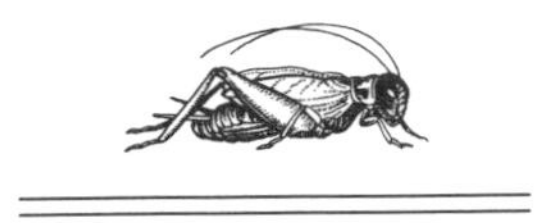

THE LEATHERBACK

Landfall for the leatherback turtle was the west coast of Costa Rica, on a steeply sloping barrier beach backed by mangrove forest. She came ashore on the high tide, a few minutes before midnight at the end of October, making a six-foot-wide track across the sand which could easily have been mistaken for that of a bulldozer but for the telltale line up the middle. I did not see her emerge from the surf, pulling herself forward with all four flippers, breathing heavily and grunting as she struggled with the unfamiliar sand that scratched her soft skin. By the time I encountered her track she was already in the dry sand above the high tide line, turned so that she was facing the dark sea, the white breakers, and the lightning that flickered on the horizon. With simultaneous sweeps of her great front flippers, she had thrown sand over her back, and now she lay still, motionless except for her much smaller hind flippers methodically excavating a hole for her eggs.

A dozen of us clustered around her barrel-shaped body, hesitantly reaching out to touch the black carapace with its white flecks and texture of hard rubber, running our fingers down one of the seven longitudinal ribs that run from back to front, discovering how soft was the skin covering her neck and flippers. Crouched in the sand beside eight hundred pounds of turtle, this boulder on a rockless beach, we told each other that leatherbacks could be twice as heavy, and dangerously pugnacious when met at sea. But once they are nesting they are imperturbable. Even when we shone our flashlights directly in her eyes, from which hung ribbons of sandy tears, she did not move her heavy blunt head nor stir her massive forequarters. The only sign that she was aware anything was happening behind her was heavy sighs, or moans.

In twenty minutes her hind flippers had scooped out the nest cavity and enlarged it at the bottom. Then, spreading her flippers so as to partially cover the hole, she lowered her stubby tail down into it and began to lay about seventy-five eggs in sets of twos, threes, and fours, interspersed by pauses. Reaching an arm between her flippers down into the hole, I caught one of the eggs as it fell, and cradled it in my palm as I examined it with a flashlight. I need not have worried about breaking it: the two-inch-diameter white sphere had a shell as tough as a plastic milk jug, and more flexible. In sixty days it would produce a hatchling leatherback; meanwhile it was the lodestone that had brought me and the turtle together on this isolated strip of tropical coast.

The egg had also drawn others. Two Costa Rican *hueveros*—egg collectors—were standing behind her when we arrived, waiting to steal the eggs as soon as they were laid. Though the taking of turtle eggs is now illegal, they have always been a windfall of protein for coastal peoples. But that isn't the only reason eggs are collected; it isn't why eggs are sold clandestinely in bars in the capital city far inland. An egg split open, the contents dumped in a glass, seasoned with hot pepper sauce and swallowed raw, is considered an aphrodisiac. For improving a man's sex life, turtle eggs bring twenty-five cents a dozen.

These poachers were probably from nearby; we, on the other hand, had come thousands of miles. Who knows how far the turtle had come. Who had the greatest claim to the eggs? Though leatherbacks are the least endangered of the sea turtles—the adults being almost universally considered inedible because of their exceptionally oily flesh—future generations depend on the survival of eggs. We had come specifically to see turtles laying eggs, contributing to the local economy for the privilege. Some thought we should make a contribution to the poachers, paying them to leave the eggs alone. Others, citing the law, wanted to become vigilantes. Cooler heads pointed out that this was not the way foreigners should behave. It was as though all the issues of modern conservation had come up on the beach with the turtle: the needs of endangered species, the rights of indigenous peoples, the pressures of tourism. In the end, the matter was moot. We milled about admiring the turtle long enough that the *hueveros* got tired of waiting and went away, in search, no doubt, of other turtles.

The turtle finished laying and began filling the nest hole with sand, packing it carefully with her flippers. When she was done, she swept the sand forcefully with her front flippers, making the beach sound like a rug being beaten, and scattering sand over the whole area. Then she started back toward the light of the sea.

She had completed her mission; she could now return to *terra infirma,* roaming the open ocean, seeking out the edges of currents and their eddies to hunt jellyfish at unfathomed depths. In a sense it didn't matter whether the poachers came back and probed the sand on the beach with a stick to find the nest chamber, or whether late rains chilled the sand so much that the

eggs failed to hatch. Only one or two of the thousands of eggs laid in her lifetime will become an adult leatherback, so in the larger scale of things a single clutch is not all that important. She did her best to lay it, we did our best to protect it—on the chance that one of the eggs was the one destined to provide her successor. Mankind's efforts to protect sea turtles are a recent development. From a turtle's point of view, the exploitation we seek to stop is relatively recent, too. Turtles laid eggs on this beach aeons before man arrived. From this perspective, we humans, poachers and protectors alike, seem inconsequential—one species for company one year, another the next. The only constant is the nesting.

Less than two hours after she emerged from the sea, the turtle was back at the water's edge. I watched as the first wave broke over her back and washed away the sand, leaving her black and glistening. The next wave picked her up and washed her shorewards, but then a wave went over her and she dug her flippers under it and was gone. I stood there in the wet sand, looking out at the empty sea and the starry sky and wished her, and us, Godspeed.

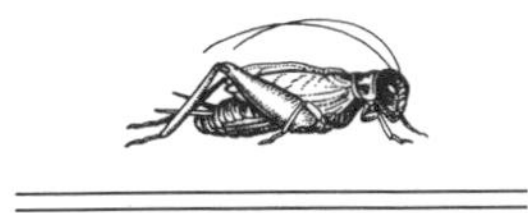

THE BOATHOUSE

The boathouse has weathered a hundred years. Some of the mossy wooden shingles have been replaced. The gray trim has been periodically repainted. But there is still an air of steam locomotives and icehouses about the place, as if their ghosts waited beneath the adjacent hemlock trees, watching from the cool needle-strewn ledges. Inside the boathouse, the one big room upstairs opens onto a broad, chair-studded porch above the lake. On the wall opposite the fieldstone fireplace hangs a mounted saltwater sailfish—the sort of wit that goes with the "No Swimming" sign in the bathroom. Downstairs, lifejackets, oars, and paddles surround a Ping Pong table. Out on the floating wooden dock, a canoe lies upside down.

The wedding guests have come from the service on the other side of town. Lest the tall, white church seem too formal, the bride and groom ushered guests into the pews and a bagpiper replaced the organ. Now laughter and strawberries fill the boathouse. Plates of new potatoes, green beans, and other early summer produce are being set out. The men take off their jackets and begin rolling up their sleeves.

There is a special quality to the hours after a wedding. It is as if the speaking of the old vows had somehow caught the pendulum of the clock at the top of its swing. Time stops. As long as the gathering holds, as long as there are plates to refill and glasses to lift to the new couple, nothing can intervene. Stories of marriages past and babies expected crisscross the tables. And with them, too, wonder at this boathouse, on whose frame the passing century seems to have left so little mark.

In this same town, the two-hour parking spaces along Main Street are filled every weekend. The New Yorkers who once came by train, on the tracks that carried away blocks of marble for building the Brooklyn Bridge, now come by car. They can't stay the whole summer anymore. But there are plenty of red geraniums and rocking chairs on the inn's front porch, and if they ignore the flash of tangerine maillots by the pool out back and concentrate instead on shopping for country curtains, they can be back in the city by Sunday night.

All over town, accommodations have been made for the crowds, but not here in the boathouse. Its charter states that regardless of any change in fortunes, there are never to be more than one hundred members. The strength of the building lies not in the oak of its beams or the rock of its foundation, but in the simple limits on its use.

However provident this provision, it raises the issue of inheritance. How many children did the founders have, and how many in turn did they have? At first guess, far too many to squeeze within these shingled walls. And yet things have been changing even in this room. Families are growing smaller. It isn't hard to see that the children who learned to swim and tie knots and sail at this boathouse could also have learned something about overloaded boats, about the merits of ample freeboard.

The number of children the bride and groom should plan is not a matter for outsiders, even for the best of friends. But all of us have a share in the world's crowding. More than our share, if you consider our standard of living. Casting about for the right wedding present for these two, we have settled on the offer to celebrate any child as our own. This is not a new gift. We grew up with it, too, but the years have left it untarnished.

By the time supper is over, the sun has set. Several

young girls gather at the dock's edge with a fishing rod. Another tucks up the skirt of her strapless dress and goes wading in the twilight. Upstairs the music is turned up for dancing, but most of us sit downstairs by the water's edge with the chorus of frogs and the citronella candles. Doctor, cook, painter, poet, we hold the world steady in our hands.

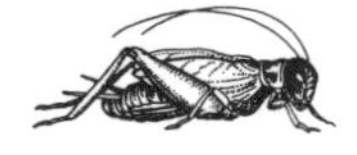